Hoverflies of Northwest Europe

Identification keys to the Syrphidae

M.P. van Veen

KNNV Publishing

CONTENTS

FOREWORD

In 1981 when it first appeared, Volkert van der Goot's book *De zweefvliegen van Noordwest - Europa en Europees Rusland, in het bijzonder van de Benelux* was arguably the most comprehensive and reliable set of keys available for the identification of Western European syrphids. However since 1981 there has been a significant increase in our knowledge of the European syrphid fauna, resulting in the publication of a succession of revisions of the European species belonging to different genera. More than 120 European syrphid species have been described as new to science since Van der Goot's volume was published.

Mark van Veen's keys build on the foundation provided by Van der Goot's book, incorporating the results of recent generic revisions, providing new keys where necessary and expanding the geographic coverage of others. These generic revisions have been published in various journals and languages, and their results have not all previously been put together in one publication. For the first time since Sack's volume in the monumental *Die Fliegen der Palaerktischen Region* series was produced in the first half of the 20th century, we have, in Van Veen's work, a set of keys that deals comprehensively with the known syrphid fauna of Northern and temperate parts of Europe. It is also the first time that such a comprehensive work on the identification of European syrphids has appeared in English. The book covers 500 species - nearly two-thirds of the known European syrphid fauna.

The upsurge in general interest in syrphids currently manifest in various parts of the continent, evidenced by the first-ever international syrphid workshop held in Stuttgart in 2001 and by its well-attended successor in Alicante in 2003, will ensure these keys are used. Their production is also particularly timely, given the efforts now being made to employ syrphids as tools in environmental assessment and interpretation, for which the availability of reliable taxonomic literature is a necessity.

Martin C.D. Speight
Invertebrate Conservation Section,
Research Branch,
National Parks & Wildlife Service,
Dublin, Ireland.

PREFACE TO THE SECOND EDITION

It was six years ago when I was working to finish this identification book on Northwest European hoverflies. I never imagined a second print. There is much interest in hoverflies and the Germans were working on their identification keys. Recently, Bartsch et al. (2009a, 2009b) published their books on the Swedish hoverflies including all Nordic species. When the publisher asked me to prepare a second edition, I wondered if there is a need to update my book. I decided there is.

In the update, I did not change the genus names nor the positioning of species in genera. Therefore all species are still in the original genus-keys, but I will remark changes in the text of the genus.

Mark van Veen
Zeist, 1 april 2010

PREFACE

When you venture out for a walk it's easy to see hoverflies flying along forest edges and near waterways. Some hover in the air defending their air space and chasing away any insects that enter it, while others visit flowers. Many have distinctive colours that catch the eye, for example, the bright yellow on black *Xanthogramma* or the red-tipped *Blera*.

Numerous species occur in residential areas. My garden in the centre of The Netherlands has attracted 49 species of hoverfly in six years, 32 of which are regular visitors. Many of these visit the flowers, while others dwell in the garden shrubs and herbs. In late summer, my flowering *Buddleia* attracts migratory species such as *Volucella zonaria* and *Eupeodes latifasciatus*. Some of the species breed in the garden and I suspect *Merodon equestris* have been eating my bulbs.

Once these insects have caught your attention, you become curious about their identity. Knowing a hoverfly's name often reveals its life history and tells you which species are related. The identity of a hoverfly in the hand can be determined by using an identification key. Van der Goot (1981) covered most of the species of Northwest Europe with his translation and elaboration of Stackelberg's (1970) identification keys. This work was the start of much interest in hoverflies. Stubbs and Falk (1983, 2002; Great Britain), Torp (1984; Denmark) and Verlinden (1991, 1994; Belgium) followed with keys for their own countries. But although each of these countries is part of Northwest Europe; none of the keys covered the whole range of species found throughout the region.

Since 1981 a good number of species have been added to the fauna of Northwest Europe. These species have recently been described or discovered in this region. Various keys to individual genera, or parts of a genus, have been published for the new species. For example, *Sphaerophoria* was treated by Goeldlin de Tiefeneau (1989, 1991), *Epistrophe* by Doczkal and Schmid (1994), *Nigrocheilosia* by Barkalov and Stahls (1997), *Spilomyia* by Van Steenis (2000), the *Cheilosia proxima*-group was presented for The Netherlands by Smit, Reemer and Renema (2001) and the *Eupeodes* of Norway by Mazanek et al. (1999a, 2004). Comparisons of closely related species that denote the differences between them have also been published, for instance by Speight (1991, *Callicera*), Doczkal (1996, e.g. *Dasysyrphus* and *Paragus*), Nielsen (1997, e.g. *Helophilus* and *Sericomyia*) and Van Steenis (1998a, *Rhingia*). *Volucella*, a journal dedicated to hoverflies, has played a pivotal role in the study of hoverflies since its foundation in 1995.

The advantage of all this accumulated work is that there are now many identification keys and much information on hoverflies available. But the disadvantage is that the information has appeared in several languages and you need a good library and many references to cover it all. The present book critically includes information from the published keys and descriptions. New (parts of) keys have also been developed from studies of specimens held in the Zoological Museum of Amsterdam collection. Using this collection, each key has been elaborated and refined.

This book focuses on the identification of species and does not include taxonomic revisions nor describe new species. For example, the problems in *Pipiza* and *Chrysotoxum* have not been resolved. To aid recognition in the field, I have described the 'jizz' of some species. The term 'jizz' is used by birders to denote characteristics that help identify a species from a distance in the field. It includes typical body form, typical colour and typical behaviour, but cannot be used to identify the species with 100% certainty. The core of the book is based on the works of Van der Goot (1981) and Verlinden (1991). I thank the late V.S. van der Goot for his continuous support in studying Diptera, without which I would never have dared to start the keys. I thank L. Verlinden for his enthusiastic permission to use his drawings, which form the backbone of the book. They were first published in his book on the hoverfly fauna of Belgium, which greatly aided hoverfly study with its good descriptions and drawings of the species. P. Grootaert (KBIN, Brussels) kindly supported this by giving his permission for the use of the drawings.

The 'determinatieklapper'*, put together by A. Barendregt, M. Reemer, W. van Steenis and Th. Zeegers for The Netherlands Hoverfly Recording Scheme was important at the start of this project. They collected reprints and made original keys to make new developments available for those working in the project. It showed that even for a small country like The Netherlands, many articles were necessary in addition to Van der Goot (1981). I thank them for their permission to use their unpublished keys.

The book could not have been made without the help of many others. D. Wolff provided me with specimens of *Pipiza accola*, T. Nielsen with specimens of some arctic species and W. van Steenis with a number of Nordic species. B. Brugge and H. de Jong gave permission to study the large hoverfly collection in the Zoological Museum of Amsterdam. There I could test all of the keys against material identified by leading hoverfly experts. I thank D. Hermes, T. Nielsen, G. Pennartz, J. van Steenis and Th. Zeegers for testing and checking (parts of) keys. M. Speight and W. van Steenis checked the last draft, resulting in many valuable improvements. T. Zeegers kindly translated Russian keys on *Parasyrphus* and *Pipiza*. A. Barkalov, D. Doczkal, H. Hippa, P. Laska, A. Maibach, I. Mazanek, T. Nielsen, M. Reemer, U. Schmid, M. Speight, J.T. Smit, J. van Steenis, Ch. Thompson, J. van der Linden, J. Vockeroth, A. Vujic and B. Wakkie permitted me to use their drawings. These cover many type specimens of newly

described species and extend the drawings of Verlinden. J. Ismay kindly helped with English terms and S. Moore edited the English. W. Seijbel (KNNV publishing) was indispendable in the last phase of publishing. Finally, I thank everybody that supported me during the years of writing the identification keys.

Scope of the book

This book contains virtually all the hoverfly species of Northwest Europe as known in 2003. The Northwest Europe of this book ranges from Ireland and Great Britain in the west to the German-Polish border in the east, and from the North Pole in the north to the Loire in Northern France in the south. It excludes all mountainous areas in Central Europe. Species occurring only at higher altitude in the Vosges, Schwarzwald, Jura and the Alps are therefore not included. Used with caution, the book will be valuable further east and south too. However, the fauna of the Mediterranean, the pannonian (e.g. the puszta of Hungary), the steppes and the Balkan differs greatly from the fauna of Northwest Europe.

The patterns of distribution are described on a large scale, using Northern, Central and Southern Europe. Northern Europe contains the Arctic and Boreal zones, including the boreal coniferous forests, and extends as far south as Skåne in Southern Sweden. Central Europe covers much of the Atlantic and continental Europe, including Great Britain, Ireland, the Low Countries, Denmark, Germany and large parts of France. Southern Europe encompasses the southern parts of the Continent and the Atlantic as well as the Mediterranean. Where appropriate, Western Europe is used to describe the Atlantic in the west (Great Britain, Ireland, and the Low Countries). As many species also occur outside Northwest Europe, the descriptions of their distributions often include other parts of world.

---

* Dutch name for a binder containing identification articles

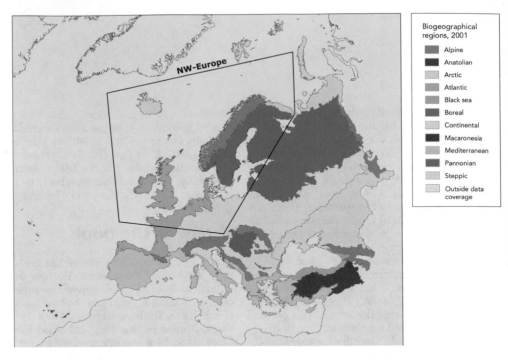

figure 1. Map of the biogeographical regions of Europe, delimiting Northwest Europe (European Environmental Agency, Copenhagen).

Some species could not be included here because their status is uncertain. For some of these species, studies are underway to clarify their status and characteristics. Females of some genera are not described as good identification criteria do not currently exist, for example, in *Paragus, Pipizella* and *Sphaerophoria*.

Mark van Veen
December 2004

INTRODUCTION

What are hoverflies?

Hoverflies come in many forms, and their appearances can vary greatly. They are particularly well known for imitating wasps, bees or bumblebees. Wearing striking yellow and black markings, a brownish honeybee-like pattern with pale body markings, or having a furry bumblebee-like appearance all help to deceive predators. Hoverflies don't just look like their models, they often behave like them. You sometimes find that the wasp which has been buzzing around some vegetation turns into a hoverfly when it lands on a leaf! This is called mimicry. Predators learn to leave stinging insects alone and will hesitate before catching anything that looks like them, giving the hoverflies time to escape. Not all hoverflies resemble stinging wasps, bees and bumblebees. A group of hoverflies (including *Xylota*) look like sawflies, which are biting wasps. Although sawflies do not sting, they are also disliked by predators, for example Brachyopa. Other hoverfly species are dull brown and black and seem to rely on cryptic colours to hide from predators. They are often found in or around vegetation. Some of these species appear quite similar to common flies of the Muscidae family and are easily overlooked.

The following paragraphs introduce the different hoverfly body forms and show how the flies look. This will help with 'on the spot' detection of species, by demonstrating what might be a hoverfly (and what might not).

Wasp mimics. These hoverflies copy the yellow and black body colouration of wasps (Photo 1, 2). Most also bear some resemblance to wasps in terms of colour pattern, form or behaviour. For example, *Eupeodes* and *Syrphus* have yellow and black stripes, but their body form and behaviour are quite different from that of wasps. They have short antennae, a broad flattened body and hover. Their appearance may still be convincing for laymen; once I was forbidden to pick up a 'wasp' that was clearly a *Syrphus*!

A number of hoverflies are excellent mimics of wasps, for example, *Temnostoma, Ceriana* and *Spilomyia*. They resemble wasps in many ways. *Ceriana* has developed long antennae on a protuberance on its head to mimic the long antennae of wasps. *Temnostoma* has short antennae, but waves its blackened forelegs in front of its head. The bodies of *Spilomyia* and *Temnostoma* are broad and rounded, just like those of social wasps. In flight, they

photo 1. The wasp mimic *Chrysotoxum bicincta* (N. Schonewille)

photo 2. A real wasp; note the folded wings, kidney-shaped eyes, and long antennae (N. Schonewille).

display the same erratic flight pattern as wasps, which is quite convincing because their quick movements help to hide their true body forms. For the skilled observer, these hoverflies betray themselves when they settle. They sit still in one place with their wings unfolded, when real wasps would begin to walk around and social wasps would also fold their wings. For me the best wasp mimic is *Sphecomyia*, a rare boreal species, which remains a convincing wasp mimic even when dead and pinned as part of a collection.

There are also some small, elongate hoverflies that exhibit yellow or grey spots or bands on a black abdomen. I find difficult to believe that they are mimicking wasps, even the small solitary wasps (Photo 3). These hoverflies include *Melanostoma, Platycheirus, Chamaesyrphus* and *Neoascia*. They mostly dwell in herbs and shrubby vegetation and are inconspicuous. Some can only be detected by sweeping the vegetation with a net. I do not know if they benefit from their body colouration or not.

Bumblebee mimics. Some hoverflies are covered in long, dense hairs; these are the bumblebee mimics (photo 4 and 5). They have different colour patterns that resemble different species of bumblebees. Popular bumblebee models include *Pyrobombus lapidarius* (black with red tip), *Pyrobombus hypnorum* (pale-haired thorax, abdomen black with white tip) and the group containing *Bombus terrestris, Bombus lucorum* and *Megabombus hortorum* (photo 6)

black with yellow bands and a white abdominal tip). The latter group is the model for the *plumata* variety of the hoverfly *Volucella bombylans* (photo 5). Other colour patterns also occur; *Arctophila fulva* and *Criorhina berberina oxycanthae* are entirely yellow-haired, and resemble pale-haired bumblebees such as *Megabombus muscorum* and pale forms of *Megabombus pascuorum*. Hoverflies differ from their models as they have short antennae (none of the Northwest European bumblebee mimics have developed long antennae), relatively large heads, and display typical quiet hoverfly behaviour. One of the best bumblebee mimics is *Pocota personata* (which resembles *B. terrestris* and *B. lucorum*), since it even has a small head.

photo 4. The bumblebee mimic *Eristalis intricaria* only deceives at first glance (M. van Veen).

photo 3. *Platycheirus tarsatus*, one of the small elongate hoverflies that have yellow spots on the abdomen (M. van Veen).

photo 5. The hairy *Volucella bombylans* is a bumblebee mimic, its *plumata* form resembles bumble bees like *Bombus terrestris* (M. van Veen).

photo 6. A real bumblebee (M. van Veen).

photo 7. A real honeybee, with long antennae (M. van Veen).

A few hoverfly species, notably *Volucella bombylans* and *Merodon equestris*, exhibit different colour forms, each of which mimic a different species of bumblebee. The typical form of *V. bombylans* is black with a red tip (like *P. lapidarius*), the *plumata* form has yellow hairs on the thorax and a pale tip (like *B. terrestris*) and the *haemorrhoidalis* form has yellow hairs on the thorax and a reddish tip (like *Pyrobombus pratorum* and several montane bumblebees).

Bee mimics. Large brown bees like the honeybee *Apis mellifera* (photo 7) and bees like *Colletes*-species and several *Andrena*-species are also popular models for hoverflies. These hoverflies have a brownish body and often show pale spots on the basal part of the abdomen. The most familiar will be the *Eristalis*-hover-flies (photo 8), most of which are large brownish hoverflies. They can be found alongside their models on flowers, but are easily distinguished from real bees by their short antennae and large eyes. Several *Merodon*-species (e.g. *M. ruficornis* and *M. natans*) and *Criorhinas* (e.g. *C. asilica*) also show this bee-like pattern. The *Merodons* occur relatively close to the ground, while the *Criorhinas* are tree dwellers.

The best bee mimics are *Callicera* and *Microdon*-hoverflies which are very convincing with their long antennae, relatively small eyes and fast flight around herbs and shrubs. *Callicera* males will hover to defend an air space and in doing

photo 8. *Eristalis pertinax* is a common hoverfly that resembles a bee (M. van Veen).

so betray their true identity, as a bee would never hover in that way. It comes as no great surprise that, in The Netherlands, *Callicera aenea* is only known from territorial males (Renema and Wakkie, 2001).

Hoverflies may also imitate smaller bees; *Eumerus-*, *Pipizella-* and *Heringia*-hover-flies look like the small *Halictus-* and *Hylaeus*-bees, especially in flight. Both these hoverflies and the bees are elongate, blackish or brownish species. Because distinct similarities in specific colouration or body form are lacking, it is not clear whether this really is mimicry or if both the hoverflies and the bees have simply evolved a cryptic body form.

Sawfly mimics. A well-defined group of hoverflies (*Xylota, Brachypalpoides* and *Chalcosyrphus*) resemble sawflies in their body form and behaviour (photo 9 and 10). These elongated hoverflies run quickly over leaves and only fly for short distances, behaving exactly like large sawflies (such as *Tenthredo atra* or *Tenthredo notha*). Some Ichneumonid wasps (including *Ichneumon suspiciosus*) which parasitise other insects also show this body form and behaviour. Stubbs and Falk (2002) noted that *X. sylvarum* is quite like an *Ectemnius* wasp, and categorised it as a wasp mimic rather than here. The imposters can be distinguished from real sawflies by their short antennae and more rounded head form.

Most *Xylota* and *Chalcosyrphus* species rarely visit flowers, instead they gather pollen that has fallen onto leaves. Notable exceptions are *Ch. nemorum* and *X. jacutorum*, which are regularly found on buttercups (*Ranunculus*) and other small flowers.

Black and brown hoverflies. A large group of hoverflies appear to rely on cryptic colouration rather than mimicry for their safety. This group includes the black *Cheilosia*s (photo 11) and the brownish *Brachyopa*s (photo 12). These hoverflies resemble common flies, especially those of the Scatophagidae, Muscidae and Tachinidae families. *Brachyopa* is notorious for behaving and appearing so much like a common fly that even experienced hoverfly catchers may pass it by without realising. Once you have the right search images, they can appear quite numerous locally. They can be distinguished from 'real' flies by head shape and the absence of strong bristles all over the body.

It's possible that the group of small hoverflies with yellow and grey spots (the group including *Platycheirus* as described in the paragraph on wasp mimics) should also be classified as cryptic and that the yellow spots help them to disappear into the patterns of shadow and light on vegetation.

photo 9. *Brachypalpoides lentus* mimics a sawfly with its elongate body form and fast pace on leaves (M. van Veen).

photo 10. A real sawfly, with long antennae (N. Schonewille).

photo 11. *Cheilosia variabilis*, one of the blackish hoverflies (M. van Veen).

photo 12. The hoverfly *Brachyopa insensilis* resembles a common fly (M. van Veen).

How to distinguish hover-flies from wasps and bees

If hoverflies are so similar to wasps, bees, bumblebees and sawflies (which all belong to the order Hymenoptera), how can they be recognised in the field? The following characteristics should help; the first four can be used on free-living insects, the last applies only to trapped specimens.

- Hymenoptera always have long, multi-segmented antennae and small, elongate compound eyes, while most hoverflies have short, three-segmented antennae and large eyes. Although some hoverflies have developed long antennae, these consist of only three segments.
- Many wasps fold their wings when they land leaving their bodies clearly visible, while hoverflies never fold their wings, so their bodies remain partly covered by their wings. However, digger wasps, bees, bumblebees and sawflies do not fold their wings.
- Hymenoptera tend to be restless when they land on flowers and leaves, they immediately want to do something. In contrast, hoverflies tend to land and stay motionless for a short time. Notable exceptions are the sawfly-mimics, which are as restless as their models.
- In general, hymenopterans do not hover. A very limited number of species prove the exception to this rule; males of the bee-genus *Melitturga* (found in Central Europe) hover close to the

ground, and a few bees, for example *Anthidium*- and *Anthophora*-bees, hover just before visiting a flower.
- In the hand, hoverflies can be seen to have two wings and two drumstick-like reduced wings (halteres) and, like all flies, they do not sting. In contrast, Hymenoptera have four wings and no halteres (beware that the hind wing is often attached to the forewing, creating the illusion of a single wing) and female bees and wasps do sting (although the smaller species are not always able to penetrate the skin).

Only the last of these characteristics is infallible, but they can all help with the recognition of hoverflies. Of course, general body form and flight style are the best characteristics for recognising hoverflies in the field. This knowledge can only be acquired by observing, not by reading.

How to distinguish hover-flies from other flies

Some hoverflies are readily identified as such by their characteristic appearance, for example, *Eristalis, Syrphus* or *Myathropa*. Others may be more difficult to differentiate from other fly families. The combination of the following typical hoverfly characteristics should be enough to determine the differences.

- The wing venation follows the pattern shown in figure 9. Hoverflies have a vena spuria (chitinized fold in the wing), a long, closed anal cell and their veins tend to follow the hind border of the wing in the tip half. The vena spuria is absent in all other fly families, although a few hoverfly genera (*Eristalinus* and *Psilota*) also have a weak or missing vena spuria.
- The arista is thread-like and implanted on the side of the third antennal segment. There are only a few genera (e.g. *Ceriana, Sphiximorpha* and *Pelecocera*). with a terminal, thick arista.
- No strong bristles on the thoracic dorsum and abdomen. Such bristles may be present on the sides of the thoracic dorsum and on the post alar lobes, e.g. in *Ferdinandea* and *Cheilosia*. However, strong bristles on the middle of the thoracic dorsum, as are found in *Muscidae*, or the abdominal segments, as are found in *Tachinidae*, are absent.

Where are hoverflies found?

A wide variety of habitats provide suitable conditions for hoverflies and their larvae. The main ecosystems are forests and woodlands, marshes, peatlands and bogs, damp grasslands, xerothermic grasslands, and gardens and residential areas. All of these systems have their own characteristic hoverfly species, largely because the larvae are limited to specific microhabitats within an ecosystem. Adults are often attracted to the flower-rich fringes of forests, marshes and grasslands. Species with a limited ability to disperse visit flowers in the immediate vicinity of the larval microhabitats. Those with greater powers of dispersal spread throughout the surrounding areas and are found far from their site of development (see Ssymank, 2001, for details).

Forests and woodlands

Forests and woodlands provide diverse habitat for hoverflies. They typically have three to four layers of vegetation: the moss, herb, shrub and tree layers. Variation in the cover provided by the different layers causes varying light and shade conditions in the forest. In addition, forests and woodlands provide plenty of dead organic material in the form of dead wood and fallen leaves as food for some of the hoverfly larvae.

The herb, shrub and tree layers structure the forest vertically, providing hoverfly habitat at different heights. The trees provide habitat for larvae that prey on arboreal aphids and caterpillars. For instance, *Parasyrphus punctulatus* larvae prey on aphids that live on the branches of coniferous trees. The larvae have longitudinal stripes on their bodies (see Rotheray, 1993), camouflaging them amongst the needles. The extent to which adult hoverflies depend on the tree layer is not known, as the upper canopy is rarely studied. Some adult hoverflies, such as *Callicera* and *Myolepta*, are reputed to live in the tree layer and to descend only rarely. That may explain why the adults are seldom seen, although the larvae can be found with relative ease if you search the rot holes where they live.

The shrub and herb layers provide habitat for other species, such as those that prey on aphids that live on herbaceous plants. The hoverflies *Platycheirus albimanus* and *P. scutatus* are commonly found on shrubs and low-growing forest plants. Their larvae mainly feed on aphids in the herb and shrub layer. The adults also dwell in these vegetation layers.

The varying light and shade conditions modify the forest habitat in many ways. In open areas and at the edge of the forest, more light reaches the forest floor. In these well-lit areas lush herbaceous vegetation can grow, including flowering plants. The sun's rays also heat these areas, providing relatively warm and dry spots. Such sites are preferred by hoverflies like *Temnostoma apiforme*, *Myathropa florea* and *Epistrophe eligans*. They bask in the sun and forage on the flowers. In contrast, a dense canopy creates shady, damp conditions. These conditions are preferred by species from the genera *Baccha*, *Sphegina* and *Brachyopa*. Apparently, these species prefer the colder microclimate that such sites provide.

Decomposing organic material provides important microhabitats for hoverfly larvae. Dead wood and rot holes in trees are a characteristic part of forest and woodland habitats and all of the hoverfly species whose larvae feed on decaying wood are typically seen as woodland species. In different phases of the decay process, different hoverfly larvae feed on the decaying wood, as described in the section on the early stages (p. 17).

In deciduous forests, fallen and decaying leaves provide an additional source of decomposing material, and are used by species like *Xylota segnis* and *Syritta pipiens*.

Some hoverflies are dependent on specific forest plants that their larvae feed on. For example, the larvae of *Portevinia maculata* and *Cheilosia fasciata* both feed on *Allium ursinum*. These hoverflies can be very abundant in sites where *A. ursinum* grows en masse. Other species, like those from the genus *Brachyopa*, feed on sap runs and depend on the deciduous trees that provide these sap runs. Beech (*Fagus*), oak (*Quercus*), elm (*Ulmus*) and poplar (*Populus*) are all important as sap run trees.

Marshes, peatlands and bogs

Marshes, peatlands and bogs provide important wetland habitats for hoverflies. They can contain areas of open water, water with emergent vegetation, peat-forming moss, damp grassland and marsh forest. Some sites are rich in flowers, particularly those along banks and in damp grasslands. As succession occurs, transitions between the different wetland environments create new habitat combinations. All these different environments provide ample habitat for both larval and adult hoverflies. Even brackish marshes are inhabited by hoverflies, in particular *Lejops vittata* and *Eristalinus aeneus*.

All aquatic hoverfly larvae depend on air for their oxygen supply. Rat-tailed maggots, such as *Eristalis* and *Parhelophilus*, retrieve air via a long anal segment that they attach to the water surface. Other larvae, such as *Melanogaster hirtella*, have a chitinous anal segment, which they use to tap air from the air channels in plant roots. The aquatic larvae of some abundant hoverflies, such as *Eristalis pertinax* and other common *Eristalis*, *Eristalinus sepulchralis*, *Helophilus trivittatus* and *H. pendulus* can survive in a wide range of waterways. But other species are much less tolerant. For example, *Anasimyia lunulata*, *Parhelophilus consimilis* and *Eristalis cryptarum* are restricted to mesotrophic peat bogs with good water quality.

Some larvae live underwater, along the roots and shoots of emergent plants and amongst the accumulated decaying vegetation. Saprophagous hoverfly larvae can be found filtering particles of detritus and bacteria from the water.

Semi-aquatic conditions occur along banks and in peat bogs. Here, decaying leaves and other vegetation build up, forming suitable habitats for *Lejogaster*, *Orthonevra* and *Chrysogaster* larvae. The adult hoverflies are attracted to the vegetation and abundant flowers on the banks. *Platycheirus fulviventris*, *Anasimyia lineata* and *Parhelophilis consimilis* are species that fly close to the waterline. *Platycheirus* are often numerous on the flowering heads of sedges (*Carex* species). The others rest on leaves and visit flowers of plants such as *Potentilla palustris* and *Cicuta virosa* which flower close to the water.

Some hoverfly larvae depend on particular marsh plants or on the aphids living on marsh plants. For example, a number of *Cheilosia* larvae feed on the thistle *Cirsium palustre*. It is the preferred food of *C. fraterna*, a common marsh *Cheilosia*. *Cheilosia albitarsis* larvae feed on the roots of buttercups (*Ranunculus*) and the adults can be abundant in marshes.

A number of predatory hoverfly larvae, especially some in the genus *Platycheirus*, specialise on aphids that feed on marsh plants. For instance, *P. perpallidus* and *P. immarginatus* larvae feed on aphids (*Thrispaphis cyperi*, *Subsaltusaphis rossneri*) that live on sedges (*Carex*) and cattails (*Typha*). Not surprisingly, the adults are found in wetland habitats.

Meadows and humid grasslands

Meadows and damp grasslands are home to a number of hoverflies that depend on herbaceous vegetation and grasses. The transition from marshes to meadows is gradual, and a combination of grasslands and open water occurs in both habitats. However while meadows and grasslands tend to be managed for agriculture, most marshes are not.

The standing vegetation provides a microhabitat for both larval and adult hoverflies. *Sphaerophoria*, *Platycheirus* and *Melanostoma* are typically found in meadows; their larvae prey on the aphids that live on herbs, sedges and grasses. *Platycheirus* and *Melanostoma* larvae also live in the soil, where they are reputed to feed on decaying vegetation (see Rotheray, 1993). Larvae of *Sphaerophoria* prefer more ruderal vegetation, for example, the meadow fringes where the adult hoverflies dwell in the vegetation. *Melanostoma* and *Platycheirus* adults feed on grass and sedge pollens; often they can only be detected by sweeping a net through the vegetation. *Sphaerophoria* feeds on flowers with shallow nectaries. Meadows and grasslands also provide a second microhabitat for larval hoverflies in the form of decaying organic material. Species such as *Syritta* and *Neoascia* breed in moist areas where decaying vegetation accumulates. The specialist *Rhingia* may be abundant in dung. *Rhingia* adults have an extremely long tongue and prefer flowers with deep nectaries, such as *Lamium*.

Aquatic microhabitats suitable for the saprophagous larvae of *Eristalis*, *Helophilus* and *Anasimyia* occur in places where ditches or streams cut through the meadows. The adults of these species can be very abundant on meadow flowers. These flowers will also attract mobile hoverflies, including many *Eristalis* and *Helophilus*, from quite a distance.

Xerothermic grasslands and rocks

Xerothermic grasslands are found in dry, hot habitats, areas usually poor in organic material. They occur both in sandy areas such as heaths and dunes and on rocky ground. Temperatures are determined by the intensity of the sun's radiation which depends on the angle and direction of the exposure to the sun, as well as by the type of substrate and the extent of barren ground. Southward facing slopes catch more sun and form hot spots in the landscape. Sandy and chalky areas with much bare soil will reach higher temperatures than vegetated parts. This habitat contains species specially adapted to a xerothermic environment (e.g. a number of *Merodon* and *Eumerus* species) and southern species that are found at the northern edge of their range in such warm spots, for instance *Paragus quadrifasciatus* on chalk hills in The Netherlands.

Within the genera *Merodon* and *Eumerus*, *M. rufus*, *M. aeneus*, *E. sabulonum*, *E. tarsalis* and *E. tricolor* are good examples of typical xerothermic species. Their larvae feed on the bulbs of the monocotyledonous species found in xerothermic grasslands, such as *Allium, Anthericum* and *Scilla*. The adults fly erratically and close to the ground and bask on stones and bare soil. They visit flowers of *Anthericum, Thymus, Geranium* and other plants.

Pipizella and *Paragus* larvae feed on root aphids that live underground as well as, for *Paragus*, the aphids that live on herbs (Roder, 1990; Rotheray, 1993; Barkemeyer, 1994). In xerothermic areas, root aphids provide a food source underground, where extreme temperatures are buffered. The adult flies fly close to the ground. *Pipizella* fly like small solitary bees in a quick, erratic way. *Paragus* are small hoverflies, that rest on the ground and in the vegetation and hover regularly. Some other species also prefer xerothermic environments. *Chamaesyrphus lusitanicus* is a species that prefers heaths, particularly those near pines (Roder, 1990). It is not known where their larvae live. *Volucella zonaria* and *V. inanis* are southern species, that survive in warm places in the north. Their larvae live in the nests of wasps, including *Vespa crabro*. *V. zonaria* is expanding its range northward, as is *V. crabro*, possibly in response to global warming.

Gardens and residential areas

Residential areas contain many green spaces, some small (gardens and green roadsides), and some large (parks). These spaces have the characteristics of all kinds of different natural environments. Areas of trees and shrubs are interspersed with grasslands and warm rocky areas (in the form of bricks and streets).

Gardens contain many species. For example, in just six years in my small garden I have seen 49 different species, 15% of the Dutch hoverfly fauna. Schmid (1996) mentions 60 species in his garden in Southern Germany. Typical garden hoverflies include species that dwell in shrubs and herbs (e.g. *Platycheirus albimanus, Meliscaeva auricollis, Sphaerophoria scripta*) and species that depend on specific garden plants (e.g. *Cheilosia caerulescens* which feeds on *Sedum* and *Sempervivum* and is a garden hoverfly in The Netherlands, or *Merodon equestris* larvae which feed on the carefully planted bulbs). These species are supplemented with opportunistic ones such as *Episyrphus balteatus* and *Scaeva pyrastri* that breed almost everywhere. In addition, garden flowers attract mobile hoverflies, such as *Eristalis* and *Helophilus* that do not breed in the garden but are frequent visitors. The flowers may also entice migratory hoverflies. Many sightings of the attractive *Volucella zonaria* are from residential areas.

Parks and larger green areas contain a wide range of hoverflies, particularly when they are more established. Such sites have the highest species richness of all residential areas. For example, parks in Oldenburg contain 80% of all species ever recorded in the city, while 23% of the city's species only occur within the parks (Barkemeyer, 1997). Weighted by individuals, species from open areas and forests dominate in Oldenburg. Schmid (1996) recorded many species in Stuttgart's Rosenstein Park including those characteristic of old trees and the rot holes they contain, such as *Sphiximorpha subsessilis*.

When is the best time to look for hoverflies?

Flower availability and the species these flowers attract change seasonally. The hoverfly season begins from early March to early May, depending on altitude and latitude. In the southern parts and near the sea, such as The Netherlands, it starts in March. At higher altitudes and latitudes, such as in Norway, it may start in May.

In early spring, willows (*Salix* species) and blackthorn (*Prunus spinosa*) are important. These shrubs flower early in spring and attract the first hoverfly species to emerge, such as *Cheilosia albipila*, *Melangyna barbifrons*, *Platycheirus discimanus* and *Parasyrphus punctulatus*. Most of these species have a single generation and disappear later in spring. A few species hibernate as adults and emerge on warm days in early spring. Common hibernators are *Eristalis tenax*, *Episyrphus balteatus* and *Meliscaeva auricollis*. *Eristalinus aeneus* also hibernates here along the Dutch coast and can be found on warm days in early spring.

Later in the season, around May in Central Europe, many more flowers become available and the hoverfly diversity increases. Many hoverflies appear at this time, which is one of the most diverse periods of the year. Herbs like cow parsley (*Anthriscus sylvestris*) and dandelions (*Taraxacum*) and shrubs such as hawthorn (*Crataegus* species) and rowan (*Sorbus aucuparius*) will attract many hoverflies. Some hoverflies, such as *Dasysyrphus venustus* and *Sphegina* species, will stay in the shade and look for flowers there. Others, such as *Eristalis* species, will go out into the full sun.

In summer, many flowers are available. At this time hogweed (*Heracleum sphondylium*) and other umbellifers, many composites and crucifers are important sources of nectar and pollen. On hot days, many species are more active in the morning and late afternoon, avoiding the high temperatures (and low humidity) of midday. A few hoverflies, for example *Cheilosia impressa*, appear for the first time in summer. Summer is also the time when the second generation of many species emerges; they are the offspring of the spring generation. Species like *Episyrphus balteatus*, *Eupeodes corollae* and *Scaeva pyrastri* become abundant in this period.

In late summer, fewer plants are flowering and the number of active hoverfly species decreases. Active species can be found in areas with flowers and sun-warmed leaves. Some crucifers still flower abundantly, for example *Brassica* and *Rapistrum* species, which grow both in cultivation and as escapees along road verges. Ivy (*Hedera helix*) flowers now, attracting many species in and around woodlands. Depending on the habitat, species like *Eristalis pertinax*, *Didea fasciata*, *Pyrophaena granditarsa* and *Helophilus trivittatus* are attracted. Some of these species are present until winter sets in. Females of *Eristalis tenax*, *Eristalinus aeneus* and *Episyrpus balteatus* will even seek shelter and overwinter. The first rays of February sun will awaken them.

Collecting and preserving specimens

The main catching devices required are the human eye and an insect net. The net is made from mesh placed on a bracket of about 30-40 cm in diameter. The net should be twice as deep as it is wide (making a depth of 60-80 cm). The bracket is placed on a long handle so it can be used to catch insects out of arm's reach, like those that hover 1.5 metres above you. Insect nets are available from entomological suppliers. Catching insects in a net requires a close inspection of the local surroundings. First, hoverflies can be located visually and caught in the net. Then, patches of good habitat can be swept with the net to catch any unseen hoverflies. Both methods should be used in an area. The first focuses on the conspicuous species, the second on the inconspicuous ones.

A number of automatic catching devices are available, of which the Malaise trap and the water trap are most commonly used. A Malaise trap is a kind of tent with a single wall in the middle. Insects that enter the tent will fly up and end in a tube at the top, where they can be collected in a bottle. Malaise traps can be made or bought. Water traps are simply coloured dishes filled with water. Insects attracted to the dishes drown in the water. Add a little formalin (not too much!) to preserve the specimens for longer than a couple of hours. The best colours appear to be yellow and white.

Captured hoverflies can be preserved by drying them on a pin (figure 2). First, the specimen is killed using ethylacetate. Simply wrap some cotton-wool in filter paper and put it on the bottom of a small glass tube (do not use plastic). Add a few drops of ethylacetate and then the specimens (in this order). After a couple of minutes, the flies will be dead (takes longer for large specimens). Pin the insects quickly after death. Entomological suppliers provide special, thin insect pins; for many hoverflies size 0 or 00 will do. Puncture the fly through the thoracic dorsum and spread its wings and legs. Let the fly dry on the pin in a box with a bottom of foam plastic to stick the pin into. After a couple of days, the fly will be dry. Then make small labels, one containing the site and date of capture and person responsible for it. A second label can be used to provide habitat details (vegetation) or the specimen's scientific name. Personally, I put the name label lowest on the pin with its face down, so that the name can be read from below and is not obscured by the labels above. Put the labels on the pin below the insect and store it in a box with a foam floor and airtight lid. These boxes can be bought from entomological suppliers.

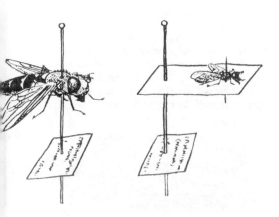

figure 2. Pinning and labelling hoverflies; on the left, large specimens are directly pinned, while on the right a small needle is used to pin it in *Polyporus*.

During field trips it is not always feasible to pin and store insects in a box. However the insects can be transported in small boxes without pinning. Choose a rigid box and use tissues to separate the layers of unpinned insects, otherwise the specimens will disintegrate. At home, soften the insects by holding them above warm water for a couple of hours (do not put them *in* the water). Then pin the specimens. A hint is to separate specimens with tissue and one or two fresh, bruised leaves of bay (*Laurus nobilis*, in the Mediterranean or Atlantic islands in the south) or cherry laurel (*Prunus laurocerasus*, a common garden tree); the water in the leaves and the evaporating chemicals keep the insects soft, while the cyanide-containing vapours will keep away mould and unwanted insects. However, too many leaves will release too much water increasing the risk of mould. Insects kept this way can be pinned at home, provided that you do not wait too long.

Early stages

Adult hoverflies are easily visible and often the only sign of hoverflies in the environment. The adults start their lives as eggs, attached to leaves, stems or dropped on the soil. The larvae that appear use a wide variety of food sources in specific microhabitats, depending on the species. Basically, there are predatory, herbivorous, fungivorous and saprophagous hoverfly larvae. Specialisation has lead to nine groups of hoverfly larvae, each with similar microhabitat and food requirements.

Eggs

The number of eggs laid per female varies from less than 100 in *Melanostoma, Platycheirus* and *Paragus*, to 4500 in polyphagous migrants like *Episyrphus balteatus* and *Scaeva selenitica*. Species with predatory larvae attach their eggs singly or in small numbers on leaves and stems close to aphid colonies and other prey species. Others deposit their eggs in or near the substrate where the larvae live, for instance *Eristalis* deposit their eggs in mud and manure. The egg stage is very brief, typically lasting less than five days, depending on the local temperature and humidity. High temperatures and high humidity accelerate development and hatching (Gilbert, 1993).

A number of strategies are used. Risk spreaders like *Episyrphus balteatus* produce many small eggs which are deposited in many patches, thus the investment per egg and per patch is low. *Eristalis tenax* also has small eggs, but has clutch sizes of up to 200 eggs. In contrast, investors like *Melanostoma scalare* deposit just one or a few eggs per patch, but the investment is much larger because they produce only a limited number of these large eggs in total. *Platycheirus* is an example of a genus that lays large eggs in small clutches. *P. fulviventris* produces a limited number of eggs and deposits them in clutches of 12 eggs on average (Dziok, 2002).

Larvae

A larvae emerges out of the egg and begins to feed. Hoverfly larvae exploit a wide range of food sources, that can be broadly classified as dead material (and associated bacteria), fungi, plants, and animals. They occur in many habitats that provide suitable food.

Larvae of species that feed on decaying material occur in all phases of the decay process. An early phase of decay starts, for example, when a tree or branch falls and the sap under the bark forms a thick, wet layer. Species that feed in this layer include *Brachyopa*, *Chalcosyrphus* and *Myathropa*. This food source disappears when the bark cracks and air dries the layer. *Temnostoma* is probably one of the first hoverflies to make use of the actual wood. It tunnels through softened but firm heartwood that has been decaying for some years (Rotheray, 1993; Speight, 2003). Once the wood is much softer and quantities of wood humus have formed, it is suitable for species like *Criorhina*, *Callicera* and *Xylota sylvarum*. These species are typically encountered in rot holes in live trees. In the later phases of decay, species that live on all kinds of decaying organic matter appear. These include *Myathropa* and *Xylota segnis*, which probably feed on the bacteria that occur in the decaying mass.

A large number of marsh species have aquatic or semi-aquatic larvae that live in decaying material and feed on detritus and bacteria. These species include the familiar *Eristalis*, *Helophilus* and their relatives. Their larvae have an elongated anal segment, which they use to retrieve air from the water surface. Because of this long 'tail', they are known as 'rat-tailed maggots'. These larvae probably filter bacteria from the water.

Melanogaster, *Chrysogaster*, *Lejogaster* and *Orthonevra* form a group with (semi-) aquatic larvae with short, chitinous anal segments (Maibach et al., 1994a, b, c). *Melanogaster hirtella* uses this segment to bore into the air channels in shoots of *Typha* and other water plants to retrieve air. Their larvae probably filter detritus and bacteria from the water.

Herbivorous larvae are confined to *Cheilosia*, *Portevinia*, *Eumerus* and *Merodon* (*Cheilosia* also contains a few fungivorous species). They specialise on different parts of the plant: *Merodon* and *Eumerus* live on the bulbs of monocotyledons, *Cheilosia* larvae tunnel stems and roots or mine

figure 3. The larva of *Eristalis tenax* live in water rich in organic material and even in liquid dung (Van der Goot, 1975).

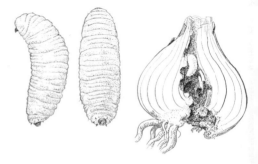

figure 4. The larvae of *Merodon equestris* live in bulbs (from Schmid (1996) after Bollow).

leaves. The maggots are rounded and legless. Using their mandibles (*Cheilosia, Portevinia*) or chitinous spines (*Merodon, Eumerus*), they crush the plant tissue and feed on the liquidized material. *Cheilosia* larvae that live in stems begin life in the stem and gradually wander to the plant roots before pupating in the surrounding soil. Species that mine leaves drop from the leaves and pupate in the soil.

Many hoverfly larvae are mono- or oligophagous and specialise on a single plant species or genus. Oligophagous species occur within *Cheilosia*, where some species have specialised on a single plant or even parts of it. A good example is *C. fasciata*, which mines the leaves of *Allium ursinum*. Stuke and Claussen (2000) describe the larval habitats of *C. canicularis, C. himantopus* and *C. orthotricha*. These larvae all feed on *Petasites hybridus*, but prefer different parts; *C. canicularis* lives in the developing shoots, *C. himantopus* in the leaf petioles and *C. ortotricha* in flowering stems. Other species are polyphagous. For example, *C. vernalis* larvae have been found feeding in *Matricaria, Sonchus, Achillea* and *Tragopogon*. The larvae of *C. pagana* have been found in *Anthriscus sylvestris* roots, but probably feed more on Umbelliferae.

Predatory hoverfly larvae mainly hunt aphids and occur within the Syrphini (e.g. *Syrphus, Sphaerophoria, Didea*), *Melanostoma, Paragus, Platycheirus, Pipiza, Pipizella, Neocnemodon* and *Triglyphus* (Rotheray, 1993). There are a large number of generalists that feed on the aphids that live on herbs, shrubs and trees. Examples are common hoverflies like *Episyrphus balteatus, Eupeodes corollae* and *Scaeva pyrastri*. Specialisation on specific vegetation layers also occurs. The larvae of *Dasysyrphus, Didea* and *Melangyna* dwell on trees and prey on tree aphids. Some occur on deciduous trees (e.g. *Dasysyrphus venustus*), others on coniferous trees (e.g. *Dasysyrphus pinastri* and *Didea alneti*). Other species prefer the herb layer, for example *Sphaerophoria, Melanostoma* and *Platycheirus* species.

Within the aphid feeders, a number of adaptations have evolved to overcome the aphids' defence systems. Aphids can produce large amounts of wax, induce galls in which to hide, or use ants to defend themselves. *Neocnemodon latitarsis* is a specialist that hunts aphids with wax excretions, *Pipiza* are specialists that hunt gall-inducing aphids.

A few species specialise on prey other than aphids, in particular caterpillars of Lepidoptera and sawflies (Symphyta) and larvae of leaf beetles (Chrysomelidae). This kind of prey feeds on leaves and is gregarious. *Trichopsomyia* larvae feed on Psyllidae larvae that form galls on *Juncus*. Finally, there is a small group of species whose larvae are associated with other insects. Larvae of *Microdon* occur in ant nests. Many *Volucella* species occur in wasp nests where they presumably feed on detritus.

How to use the book

The aim of the book is to identify hoverflies. The main part of this book is dedicated to keys to the species. It starts with a key to the genera, followed by keys to each species in a particular genus.

Using the keys

In principle, using a key is simple. You start with the first item. It contains two parts that contradict each other. You compare the specimen at hand with the characteristics mentioned and select the part that matches your specimen. It then refers you to the next item by the number of that item in the key. You go to that item and compare the contradicting characteristics in each of its parts. Finally, you arrive at the name of the genus and further on of the species.

figure 5. Larva of *Syrphus torvus*, a species that preys on aphids (Van der Goot, 1975).

Difficulties arise at two points. First, specimens of a single species are not identical but show variation. As an example, the amount of black colouration on the legs will vary. In the key, characteristics are selected that are typical for a species within its whole range of variation, aiming at zero overlap with other species. This is not always possible, leaving a limited fraction that just matches the characteristics of other species. For example, dark specimens may occur in otherwise pale species.

Secondly, it is essential that the reader correctly interprets the characteristics. A number of issues are critical. The form and placement of pale patterning is essential in keying out specimens. As often as possible, figures have been provided to illustrate the patterns. Also, determining whether parts are dusted or shiny can be difficult and may only be achieved by turning specimens in the light.

Critical reading and observation is the key to solving both points. Read all the characteristics listed and compare them with the specimen, never use a single characteristic if more than one is given. Once you arrive at a species, more characteristics are often given. Compare them with the specimen. If these do not fit, doubt the identity of the specimen and check where things may have gone wrong.

Likewise, check the characteristics of the key and the body part they apply to. For example, if the key mentions that the femur is completely pale and you observe that the very base of the leg is black, check if the black is really on the femur. The dark part may be the trochanter only. A good understanding of the body parts is essential in interpreting the characters on the side of the thorax, where many chitin plates are present and characters often apply to only one of them. To make things even more difficult, aberrant specimens occur. Sometimes anomalies can be detected by comparing left and right sides.

Examining specimens from established collections that have been identified by experts is also helpful. These provide a base for comparison with specimens that you have collected yourself. They may be used to interpret the key for any items where you have doubts, and to check characteristics and variability.

THE HOVERFLY BODY

Body

A hoverfly's body consists of a head, a thorax and an abdomen (figure 6). This section provides descriptions and names for the main structural features associated with these body parts.

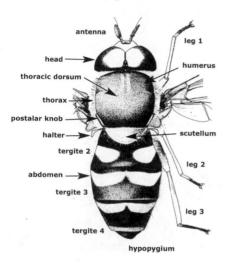

figure 6. Basic body parts of a hoverfly.

Head

Much of the head (figure 7) is occupied by the compound eyes, which are usually brownish or reddish in colour. In females, the compound eyes are separated by a wide gap (dichoptic), while in most males the eyes meet on the upper side of the head (holoptic). This makes the compound eyes a useful characteristic for distinguishing males from females in the field. In some genera the males' eyes are also separate, for example, in *Microdon, Neoascia* and *Helophilus*. Males can then be recognised by the globular genitalia at the tip of the abdomen.

Two antennae are embedded on the front of the head. The antennae consist of four parts: the first, second and third segments and the arista (implanted on the third segment). The first two segments are generally small, the third is large. The arista is hair-like. Two moon-like structures are visible just above the antennae, these are the lunula. The frons is the area between the lunula and the compound eyes. On top of the head near the thorax lie three slightly raised ocelli. Below the antennae, most species have a facial knob, which is most obvious when viewed side on. Below the facial knob is the mouth edge, visible as a rim in most species. In a limited number of species the mouth edge forms a long snout, for example, in *Rhingia*. The area below both compound eyes is called the genae. This can be a small strip, as in most species, or it may be wider.

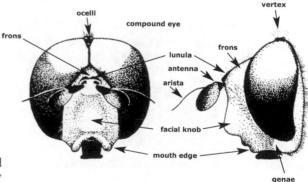

figure 7. Structural features of a hoverfly head.

Thorax

Viewed from above, the thorax consists of the thoracic dorsum and behind that, the semi-circular scutellum (figure 6). The legs and wings are attached to the thorax. The thoracic dorsum may display colour patterns, such as white stripes or yellow spots. The sides of the thorax consist of many different parts (figure 8). The presence of hairs

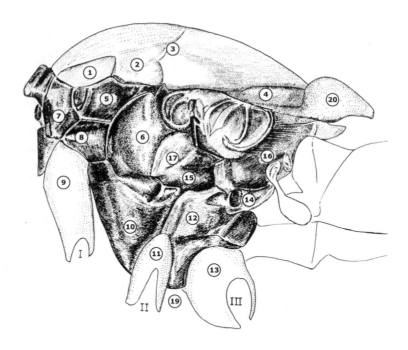

figure 8. Structural features of the sides of the thorax. Numbers correspond to those listed in table 1.

Table 1. Body parts of the sides of thorax.

Number in figure 8	Name used in this book	Alternative	Dutch name (Van der Goot, 1981)
1	humerus	postpronotal lobe	schouderknobbel
2	notopleuron		
3	suture		dwarsnaad
4	postalar knob	postalar callus	achterhoeksknobbel
5	anterior anepisternum	anterior mesopleuron	voorste vlakke deel middenzijplaat
6	posterior anepisternum	posterior mesopleuron	middenzijplaat
7	proepisternum		
8	proepimeron		
9	coxa 1		
10	katepisternum	sternopleuron	borstzijplaat
11	coxa 2		
12	meron	hypopleuron	onderzijplaat
13	coxa 3		
14	metepimeron		
15	katepimeron		
16	katatergite	metapleuron	achterzijplaat
17	anterior anepimeron	pteropleuron	vleugelzijplaat
18	posterior anepimeron		
19	metasternum (below, between coxa 2 and 3)		
20	scutellum		

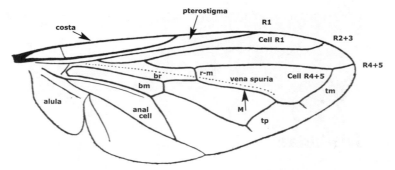

figure 9. A hoverfly wing, showing the cells and veins.

or dusting on these parts are often useful diagnostic features. There are a number of naming conventions for these parts, I large-ly follow those used by Vockeroth (1992) and Van Steenis (2000). Table 1 provides a key to figure 8, and also contains alternative English names as well as the Dutch names used in Van der Goot (1981).

Wing

The veins in hoverfly wings are another useful diagnostic feature. The series of veins near the front margin are called the radius-veins (denoted by an R in figure 9). The set in the middle are called the medius-veins (denoted by an M). Hoverflies typically have venia spuria, a vein-like fold in the wing. At the basis of the wing, at the hind margin, two wing flaps are present: the calypter and the alula. The calypter is connected to the thorax and somewhat folded. The alula follows and is separated from the remain-der of the hind margin by an incision of the hind margin. The areas of wing between the veins are called cells. The

cells may be partly or completely covered by microscopic hairs called microtrichia. These are invisible to the naked eye and only appear as a grey colour when viewed through a hand lens. Under a backlit binocular microscope microtrichia are visible as small dark dots and stripes on the wing. The extent of microtrichia on the basal cells *br* and *bm* is diagnostic for a number of species.

Legs

Each leg has five main parts: the coxa, trochanter, femur, tibia and tarsus. The tarsus consists of five segments, the first is called the metatars. The last tarsal segment has hooks and suckers that help the hov-erfly adhere to various surfaces. Through-out the keys in this book, the legs are numbered from front to back. The front pair of legs are number 1, and the hind legs are number 3. Each part of a leg is numbered accordingly, thus femur 3 is the femur of the hind leg.

Abdomen

The abdomen is made up of segments, each of which has two parts. The top or dorsal parts are called tergites (remember: **top** = **t**ergites), the lower or ventral parts are called sternites. In males, the last segment curves underneath the abdomen holds the geni-talia and is visible as a bulbous structure. Their last visible abdominal segment from above is the pregenital segment and is called the hypopygium. The form of the genitalia is often highly diagnostic and depends on the genus and species. Where necessary, parts of the genitalia are explained in the species keys. To view genitalia, it is essential that they have been unfurled from the abdomen. Place a needle between the geni-talia and the abdomen and twist the geni-talia backwards. They will pull away from the abdomen.

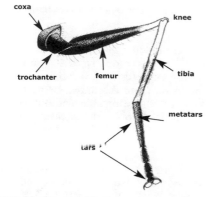

figure 10. Structural features of a hoverfly leg.

KEYS

Key to Syrphidae genera

The key to the genera uses some species characteristics that are only true of the Northwest European members of a particular genus, and is therefore restricted to this region. In other parts of the world, species occur that will not fulfil the genus characteristics of this key.

1.a. Humeri bare (figure 11); head strongly convex posteriorly and closely adpressed to thorax so that humeri are partly or entirely hidden. Male: tergite 5 visible in dorsal view **> Syrphinae** (p. 24)
1.b. Humeri haired (figure 12); head not strongly convex posteriorly, humeri clearly exposed. Male: tergite 5 not visible in dorsal view **> Milesiinae, Eristalinae** and **Microdotinae** (p. 28)
   Note: Microdontinae are sometimes regarded as a separate family, Microdontidae.

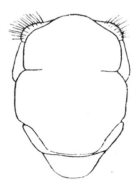

figure 12. Thoracic dorsum of Milesiinae, Eristalinae, Microdontinae (after Stackelberg).

figure 11. Thoracic dorsum of Syrphinae (after Stackelberg).

Syrphinae

Most Syrphinae are wasp mimics, but the majority are not very convincing. Only species from *Chrysotoxum*, with their stout bodies and long antennae, and *Doros*, with their petiolate bodies and dark wing front, are convincing wasp mimics, particularly when in flight. *Eriozona syrphoides* is the only bumblebee mimic. *Platycheirus* and allies are inconspicuous hoverflies that dwell in vegetation.

1.a. Antennae short and rounded, shorter than the head (figure 13) **>** 2
1.b. Antennae very elongated, longer than the head (figure 14) **> Chrysotoxum** Meigen (p. 83)
   Jizz: wasp mimics with a broad abdomen.

2.a. Face below the antennae has a facial knob and a more or less protruding mouth edge (figure 13) **>** 3
2.b. Face below the antennae protrudes forwards in a semicircle, without a separate facial knob and protruded mouth edge (figure 15) **> Paragus** Latreille (p. 155)

3.a. Tergite 2 broader than scutellum **>** 4
3.b. Tergite 2 extremely elongated and narrow, smaller than scutellum **> Baccha** Fabricius (p. 40)

4.a. Scutellum and face black, often with greenish sheen ❯ 5
4.b. Scutellum or face, normally both, largely yellow ❯ 10

5.a. Eyes bare ❯ 6
5.b. Eyes haired ❯ *Melangyna quadrimaculata* (p. 130,132)

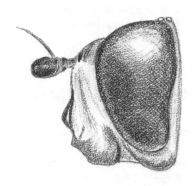

figure 13. *Parasyrphus lineola*, head of male (Verlinden).

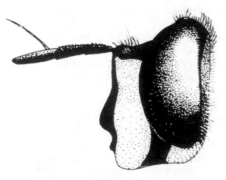

figure 14. *Chrysotoxum cautum*, head of male (after Stackelberg).

figure 15. *Paragus haemorrhous*, head (Verlinden).

6.a. Abdomen elongate, more than twice as long as wide, completely black or black with yellow, grey or reddish markings; thoracic pleurae: katepisternum with pile patches broadly separated throughout ❯ 7
6.b. Abdomen broad and flattened, less than twice as long as wide, with reddish-yellow markings; thoracic pleurae: katepisternum with pile patches joined anteriorly ❯ *Xanthandrus* Verrall (p. 224)

7.a. Abdomen: entirely black or black with subequal grey, orange or yellow pairs of spots at least at tergites 2–4; wing hyaline ❯ 8
7.b. Abdomen: either tergite 2 black on front, red on hind part, tergite 3 entirely red (male) or red in front part (female) and tergite 4 with large red spots or tergites black with only a pair of large yellow spots on tergite 3, tergite 4 only with a pair of small, faint spots; wing infuscated ❯ *Pyrophaena* Schiner (p. 192)

8.a. Metasternum entire, broad over its entire length (figure 17); metatars 1 flattened in many, but not all, species. Males of most *Platycheirus*-species: tibia 1 and/or metatars 1 broadened ❯ 9
8.b. Metasternum greatly reduced, sclerotized portion small with diamond-shaped central area (figure 16). Males: tibia 1 cylindrical, not broadened at tip, tars 1 cylindrical, as wide as tip of tibia 1. Female: tars 1 cylindrical ❯ *Melanostoma* Schiner (p. 137)

9.a. Tergites 3 and 4 with spots (if present) rectangular or trapezoid, often rounded in their corners, coloured yellow, grey or blueish. Males: tibia and/or tars 1 flattened and broadened, with special bristles or pile, if both tibia 1 and metatars 1 cylindrical, then femur 1 with a long ?-shaped hair near the tip (*P. ambiguus* and relatives). Female: tars 1 dorsoventrally flattened in most species ❯ *Platycheirus* Lepeletier & Serville (p. 171)
Note: including *Pachysphyria*.
9.b. Tergites 3 and 4: spots always present, narrowly triangular, pointed at their median corner. Males and females: tibia and metatars 1 cylindrical, femur 1 without ?-shaped hair; ❯ *Melangyna (cingulata)* (p. 129)

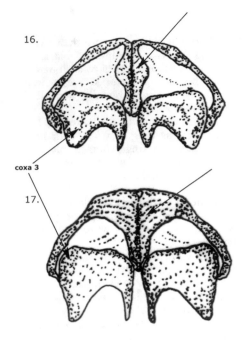

16.

coxa 3

17.

figure 16. *Melanostoma*, thorax, ventral view.
figure 17. *Platycheirus*, thorax, ventral view
(after Thompson and Rotheray,1998).

10.a. Anterior anepisternum pilose, at
least posterodorsally **>** 11
10.b. Anterior anepisternum bare **>** 13

11.a. Metasternum bare; tergites black
with a yellow band or yellow spots **>** 12
11.b. Metasternum haired; tergites orange
(but often darkened) with a typical pattern of a black 'moustache' on the front
1/2 and a black band along the hind
margin **>** *Episyrphus* Matsumura (p. 95)

12.a. Eyes bare or haired; wing: hind margin without minute, closely spaced,
black maculae **>** *Parasyrphus* Matsumura (p. 157)
12.b. Eyes bare; wing: hind margin with
minute, closely spaced, black maculae
on hind margin (figure 18) **>**
Episyrphus (Meliscaeva) Frey (p. 97)

13.a. Thoracic dorsum: lateral margins
with a sharply-defined yellow band, at
least from humerus to suture, often
extending to postalar knob **>** 14
13.b. Thoracic dorsum: lateral margins
without sharply-defined band, black or
at most a dull yellow band, not sharply
contrasting with median parts **>** 17

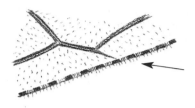

figure 18. *Melangyna cincta*, detail of the
hind margin of the wing (Verlinden).

14.a. Tergites 2-4 strongly marginated,
with marginal ridge **>** 15
14.b. Tergites 2-4 not marginated, without marginal ridge **>** 16

15.a. Thoracic pleurae with yellow spots
> *Xanthogramma* Schiner (p. 225)
15.b. Thoracic pleurae without yellow
spots **>** *Epistrophe* Walker partim (p. 93)

16.a. Wing hyaline; abdomen elongate,
with parallel margins, or weakly petiolate; smaller: 12 mm or less. **>** *Sphaerophoria* Lepeletier & Serville (p. 200)
16.b. Wing darkly infuscated at front margin; abdomen strongly petiolate, narrowed at the hind margin of tergite 2;
larger: 14–16 mm. **>** *Doros* Meigen (p. 92)

17.a. Tergites 2-4: either with subequal
yellow spots or bands, at most those on
tergite 2 somewhat larger or at least tergite 2 black **>** 18
17.b. Tergites 2-4: tergite 2 with large
white or grey spots, which may merge
into a band (in rare cases tergite 2 largely black, whitish on median front), tergites 3 and 4 black or with small, linear,
white or grey spots (figure 19) **>**
Leucozona (p. 126)

18.a. Tergites 3 and 4: with pale markings,
not densely haired **>** 19
18.b. Tergites 3 and 4: black, densely
haired **>** *Eriozona* Schiner partim (p. 98)

19.a. Wing: vein R4+5 distinctly sinuate
(figure 20) **>** 20
19.b. Wing: vein R4+5 straight or nearly
so (figure 21) **>** 23

20.a. Metasternum bare **>** 21
20.b. Metasternum pilose **>** 22

figure 19. *Leucozona laternarium*, abdomen of female (Verlinden).

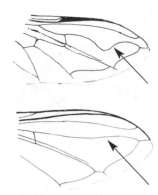

figure 20. *Didea fasciata*, tip of wing.
figure 21. *Eupeodes luniger*, tip of wing (after Stackelberg).

21.a. Eyes bare; wing: densely covered with microtrichia at least on top 2/3 ❭ *Eupeodes* Osten Sacken (p. 112)

21.b. Eyes haired; wing: hardly covered by microtrichia, with extensive bare areas in top 2/3 ❭ *Scaeva* Fabricius (p. 196)

22.a. Abdomen broadly oval, with connected spots shaped like sunglasses (figure 22) ❭ *Didea* Macquart (p. 91)

22.b. Abdomen broad, but more elongate, tergites 3 and 4 with rectangular yellow bands ❭ *Eriozona (Megasyrphus)* Schiner (p. 98)

23.a. Abdomen marginated, with well-defined marginal ridge; abdomen oval ❭ 24

23.b. Abdomen not marginated, without ridge; abdomen parallel-sided, elongate ❭ *Melangyna* Verrall (including *Meligramma*) (p. 129)

24.a. Calypter bare; thoracic dorsum dull or shiny ❭ 25

24.b. Calypter with lower lobe pilose, especially on posteromedian portion (figure 23); eyes bare or haired; thoracic dorsum dull with slight green sheen ❭ *Syrphus* Fabricius (p.215)

Note: *S. nitidifrons* may have an (almost) bare calypter, but tibia 3 black and mouth edge black, in contrast to the species following 24a.

25.a. Eyes bare or with sparse hairs ❭ 26

25.b. Eyes densely haired ❭ *Dasysyrphus* Enderlein (p. 88)

26.a. Sternites: at least sternite 3 (often all sternites) with distinct black markings; metasternum haired; abdominal marginal ridge starts on tergite 2; tergites 3 and 4: yellow markings forming spots or sinuate bands ❭ *Eupeodes* Osten Sacken (p. 112)

figure 22. *Didea intermedia*, abdomen (Verlinden).

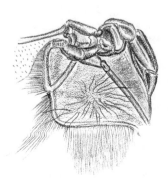

figure 23. *Syrphus*, upper side of calypter (Verlinden).

24.

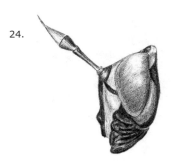

25.

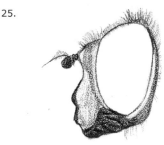

figure 24. *Spiximorpha subsessilis*, head.
figure 25. *Eristalis nemorum*, head of male
(Verlinden).

26.b. Sternites yellow, without black markings; metasternum bare or haired, if metasternum haired then abdominal marginal ridge starting at tergite 3; tergites 3 and 4: yellow markings often straight bands, sometimes forming spots❯ *Epistrophe* Walker partim (p. 93)

Milesiinae, Eristalinae and Microdontinae

The hoverflies in these subfamilies have a variety of colour patterns, including the yellow on black wasp mimics, hairy bumblebee mimics and brown bee mimics. They also include some completely black and very inconspicuous species.

1.a. Antennae longer than the head (figure 24) ❯ 2
1.a. Antennae shorter than the head (figure 25) ❯ 6

2.a. Scutellum not armed; abdomen with or without yellowish markings. If abdomen blackish, then arista on tip of 3rd segment (*Callicera*) (Milesiinae partim) ❯ 2

2.b. Scutellum armed with 2 teeth (figure 26); abdomen dull brown, black or greenish, without yellowish markings; arista implanted on basal part of 3rd antennal segment (Microdontinae) ❯ *Microdon* (p. 144)

3.a. Antenna with terminal stylus (figure 24) ❯ 4
3.b. Antenna with dorsal arista (figure 27) ❯ 5

4.a. Tergites 2-4 black with slight metallic sheen; eyes haired; face: frontal prominence absent ❯ *Callicera* Panzer (p. 48)
4.b. Tergites 2 and 3 black with yellow markings; eyes bare; face: frontal prominence absent or present ❯ *Ceriana* Rafinesque and *Sphiximorpha* Rondani (p. 49)

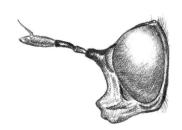

figure 26. *Microdon devius*, scutellum.
figure 27. *Psarus abdominalis*, head of male
(Verlinden).

5.a. Abdomen largely reddish; face: frontal prominence long and elongate (figure 27) ❯*Psarus* Latreille (p. 191)
5.b. Abdomen black with yellow bands; face: frontal prominence small, triangular ❯ *Sphecomyia* Latreille (p. 209)

6.a. Arista plumose, hairs longer than 3 times the arista width (figure 28) (Milesiinae) ❯ 7
6.b. Arista short haired or bare, hairs at most twice the thickness of arista (figure 29) ❯ 11

28.

29.

30.

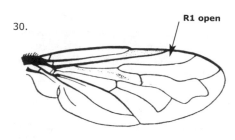

R1 open

figure 28. *Sericomyia silentis*, antenna.
figure 29. *Brachyopa panzeri*, antenna (Verlinden).
figure 30. *Eumerus sabulonum*, wing (after Van der Goot, 1981).

31.

R1 closed

32.

33.

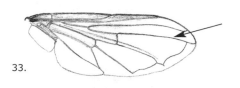

figure 31. *Volucella*, wing (after Van der Goot, 1981).
figure 32. *Eristalis,* wing.
figure 33. *Cheilosia,* wing (Verlinden).

7.a. Cell R1 open (figure 30) ❯ 8
7.b. Cell R1 closed (figure 31) ❯
  Volucella Geoffroy (p. 223)

8.a. Abdomen reddish to orange-brown, with short, sparse hairs and without yellow markings ❯ 9
8.b. Abdomen either brown with long, dense hairs or black with yellow markings ❯ 10

9.a. Wing: tm ends on R4+5 in a straight angle; abdomen more elongate; larger species: 10-12 mm ❯
  Hammerschmidtia Schummel (p. 119)
9.b. Wing: tm ends on R4+5 in a sharp angle; abdomen shorter, more rounded; smaller: 6-9 mm ❯ *Brachyopa* Meigen partim (p. 40)

10.a. Abdomen without yellow markings, long-haired ❯ *Arctophila* Schiner (p. 39)
10.b. Abdomen with yellow markings, short-haired ❯ *Sericomyia* Meigen (p. 199)

11.a. Wing: vein R4+5 strongly sinuate, arcs into underlying cell (figure 32) ❯ 12
11.b. Wing: vein R4+5 straight or almost so, forms a straight upper margin of the underlying cell (figure 33) ❯ 24

12.b. Femur 3 without tooth or triangular plate (Eristalinae) ❯ 13
12.a. Femur 3 with large triangular plate below the tip (figure 34) or a tooth at 1/4 of the femur length before the tip (Milesiinae partim) ❯ 22

13.a. Wing: cell R1 closed (figure 35) ❯ 14
13.b. Wing: cell R1 open (figure 36) ❯ 15

14.a. Scutellum black; eyes spotted ❯
  Eristalinus Rondani (p. 99)
14.b. Scutellum brown to pale; eyes not spotted, for 1 species (*E. tenax*) with 2 bands of hair ❯ *Eristalis* Latreille (p. 100)

15.a. Thoracic dorsum partially with greyish or yellowish dust patterns, either as longitudinal stripes or dorsum yellowish dusted with large black spots ❯ 16
15.b. Thoracic dorsum densely haired or uniformly dusted ❯ *Mallota* Meigen (p. 128)

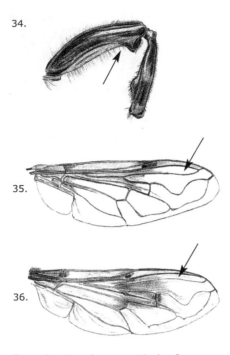

34.

35.

36.

figure 34. *Merodon equestris*, leg 3.
figure 35. *Eristalis cryptarum*, wing.
figure 36. *Mallota fuciformis*, wing
(Verlinden).

16.a. Eyes bare ❭ 17
16.b. Eyes haired ❭ 21

17.a. Antennae: reddish to black, 3rd segment longer than wide; tergites with
yellow and grey spots, at least on tergite
2 partially yellow ❭ 18
17.b. Antennae: black, 3rd segment wider
than long; tergites with longitudinal,
grey stripes ❭ **Lejops** Rondani (p. 126)

18.a. Face without bare stripe; antennae:
black or reddish–orange ❭ 19
18.b. Face with a bare, yellow or black
stripe; antennae black ❭ **Helophilus**
Meigen (p. 119)

19.a. Leg 3: at least tibia and often femur
a quarter or more pale; arista largely reddish. Males: eyes separated on frons ❭ 20
19.b. Leg 3: femur, tibia and tarsus black,
at most knees (part where femur and
tibia connect) orange; arista black.
Males: eyes meet on frons ❭ **Mesembrius**
Rondani (p. 143)

20.a. Tibia 3: with 2 dark bands, yellow in
between; ocelli: distance between hind
ocellus and eye margin larger than the
diameter of an ocellus ❭ **Anasimyia**
Schiner (p. 37)
20.b. Tibia 3: with 1 dark band on the tip
half; ocelli: distance between hind ocellus
and eye margin shorter than the diameter
of an ocellus ❭ **Parhelophilus** Girschner
partim (p. 162)

21.a. Thoracic dorsum yellowish dusted
with 3 large black spots, 2 on anterior and
1 on posterior half. Males: eyes meet on
frons ❭ **Myathropa** Rondani (p. 145)
21.b. Thoracic dorsum blackish with 2 longitudinal yellowish dust stripes. Males: eyes
do not meet on frons ❭ **Parhelophilus**
Girschner partim (p. 162)

22.a. Femur 3 with a triangular plate below
the tip; wing: cell R1 open ❭ 23
22.b. Femur 3 with a conical tooth at 1/4 of
the femur length before the tip; wing: cell
R1 closed ❭ **Milesia** Latreille (p. 145)

23.a. Face without longitudinal keel; tergites
2 and 3 black with at most a pair of triangular pale spots, which may only be extensive
on tergite 2 ❭ **Merodon** Meigen (p. 139)
23.b. Face with longitudinal keel; tergites 2
and 3 both largely red with a longitudinal
median black band ❭ **Tropidia** Meigen
partim (p. 222)

24.a. Crossvein r-m stands on top half, often
top 1/3, of discal cell (figure 37) ❭ 25
24.b. Crossvein r-m stands before the middle of discal cell (figure 38) ❭ 38

25.a. Femur 3 with a strong triangular or
conical tooth below (figure 39, figure 40)
❭ 26
25.b. Femur 3: without tooth (figure 41) ❭
27

26.a. Abdomen: black with red pattern;
femur 3: with a triangular tooth below the
tip (figure 39); face with keel ❭ **Tropidia**
Meigen partim (p. 222)
26.b. Abdomen: black with yellow bands;
femur 3: with a conical tooth below (figure 40); face without keel ❭ **Spilomyia**
Meigen (p. 213)
Note: *Milesia* has the same tooth on femur 3, but vein
R4+5 is clearly bent and it lacks yellow thoracic pattern just in front of the scutellum.

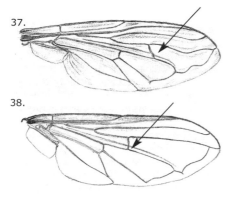

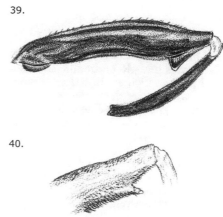

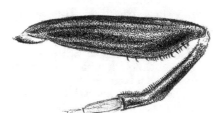

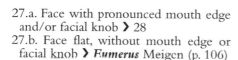

figure 37. *Tropidia scita*, wing.
figure 38. *Cheilosia albitarsis*, wing
(Verlinden).

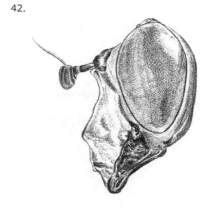

27.a. Face with pronounced mouth edge and/or facial knob **>** 28
27.b. Face flat, without mouth edge or facial knob **>** *Eumerus* Meigen (p. 106)

28.a. Face does not protrude ventrally, with clearly marked mouth edge and, sometimes, facial knob **>** 29
28.b. Face protrudes ventrally, tuberculate, genae broad (figure 42) **>** *Criorhina* Meigen (p. 86)

29.a. Abdomen covered with long erect hairs, bee or bumblebee mimics **>** 30
29.b. Abdomen either covered with short hairs or, in some species, covered with longer but adpressed hairs, that hide the background colour of the tergite, does not resemble bees or bumblebees **>** 31

30.a Abdomen broad with long hairs, in yellow and black bands, a bumblebee mimic **>** *Pocota* Lepeletier & Serville (p. 189)
30.b Abdomen elongate, with long brown hairs, bee mimic **>** *Brachypalpus* Macquart (p. 45)

31.a. Face partly yellow; abdomen copper green **>** 32
31.b. Face black, sometimes partly dusted grey or white; abdomen not copper green **>** 33

figure 39. *Tropidia scita*, femur and tibia 3
figure 40. *Spilomyia saltuum*, tip of femur 3
figure 41. *Xylota nemorum*, femur and tibia 3
figure 42. *Criorhina ranunculi*, face of male
(Verlinden).

32.a. Frons not strongly protruding at antennal implant; postalar knob and scutellum with black bristles; smaller: 8-12 mm ❯ *Ferdinandea* Rondani (p. 118)

32.b. Frons strongly protruding at antennal implant (figure 43); abdomen elongate, with copper sheen and yellow pile; postalar knob and scutellum without bristles; larger: 14-15 mm ❯ *Caliprobola* Rondani (p. 47)

33.a. Thoracic dorsum without yellow patterning at disk, at most whitish on the side margin; abdomen: black or black with grey, yellow or red spots, or a large red spot on tergites 3 and 4 ❯ 34

33.b. Thoracic dorsum with yellow patterning on the disk; abdomen: black with yellow bands ❯ *Temnostoma* Lepeletier & Serville (p. 217)

34.a. Abdomen elongate with parallel margins, black, black with yellow spots, or black with a large red spot ❯ 35

34.b. Abdomen broadly oval, either with 1st tergites black and last tergites red or completely black (figure 44) ❯ *Blera* Billberg (p. 41)

35.a. Metasternum with hairs at least as long as those on the ventral area of the katepisternum ❯ 36

35.b. Metasternum bare or almost bare (hairs much shorter than those on the ventral area of the katepisternum) ❯ 37

36.a. Femur 3: greatly enlarged; wing: almost bare on basal 2/3 ❯ *Syritta* Lepeletier & Serville (p. 215)

36.b. Femur 3: not enlarged; wing: mostly microtrichose, at most moderate bare areas on basal 1/3 ❯ *Chalcosyrphus* Curran (p. 50)

37.a. Upper and lower katepisternal hairs in patches connected across the central area of the sclerite anteriorly by scattered somewhat shorter hairs; tergite 3 entirely pale-haired and entirely orange; legs entirely black, 11-14 mm. Europe, east into Asia minor. ❯ *Brachypalpoides* Hippa

37.b. Upper and lower katepisternal hairs patches distinctly and broadly separated, the area between them entirely bare; tergite 3 reddish or black with yellowish to greyish spots; legs entirely black or with pale parts ❯ *Xylota* Meigen (p. 227)

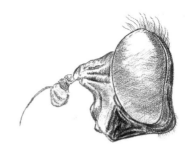

figure 43. *Caliprobola speciosa*, head of male (Verlinden).

figure 44. *Blera fallax*, habitus of male (Verlinden).

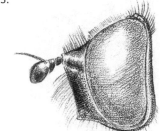

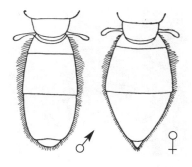

figure 48. *Triglyphus primus*, abdomen of male and female (after Stackelberg).

46.

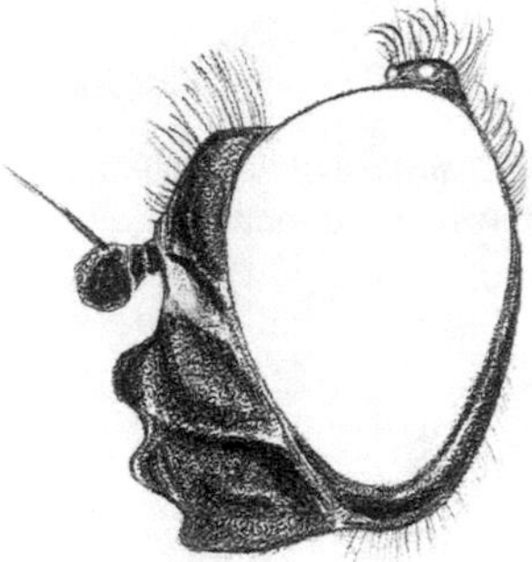

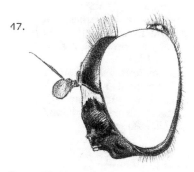

47.

figure 45. *Pipiza lugubris*, head of male
figure 46. *Melanogaster aerosa*, head of male
figure 47. *Chrysogaster coemiteriorum*, head of male (Verlinden).

38.a. Face: flat, profile straight between antennal implant and mouth edge, without projecting facial knob or mouth edge and not inflated (figure 45) ❯ 39
38.b. Face: with a projecting facial knob, mouth edge or both (figure 46) or inflated and hiding most of the facial knob and mouth edge (figure 47) ❯ 44

39.a. Abdomen: tergites 2-4 of about equal size ❯ 40
39.b. Abdomen: tergites 2 and 3 large, tergite 4 small (figure 48) ❯ *Triglyphus* Loew (p. 221)

40.a. Anterior anepisternum bare ❯ 41
40.b. Anterior anepisternum with long hairs ❯ *Trichopsomyia* Williston (p. 219)

41.a. Antennae: 3rd segment short, at most 1.5 times longer than wide ❯ 42
41.b. Antennae: 3rd segment elongate, at least twice as long as wide ❯ 43

42.a. Frons: protrudes conically forward at site of antennal implant; abdomen: black or black with yellow spots. Males: trochanter 3 without a long spine. Females: frons with dust spots (except for P. quadrimaculata, there tergites with 2 pairs of spots) ❯ *Pipiza* Fallén (p. 165)
42.b. Frons: does not protrude forward; abdomen: black. Males: trochanter 3 with a long spine. Females: frons without dust spots; tergites black ❯ *Heringia (Neocnemadon)* Rondani (p. 122)

43.a. Wing: vein tm with a notch at 1/3 from the beginning, ends in a sharp angle on vein R4+5 (figure 49) ❯ *Heringia (Heringia)* Rondani (p. 122)
43.b. Wing: vein tm with a notch halfway, ends perpendicular to vein R4+5 (figure 50) ❯ *Pipizella* Rondani (p. 169)

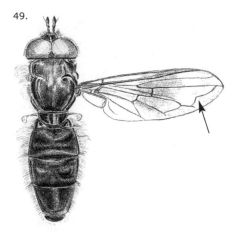

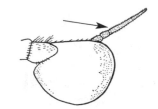

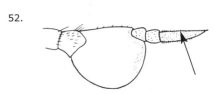

figure 49. *Heringia heringi*, male body and wing.
figure 50. *Pipizella varipes*, male body and wing (Verlinden).

figure 51. *Chamaesyrphus*, antenna
figure 52. *Pelecocera*, antenna (after Stackelberg).

44.a. Arista hair-like, placed dorsally and often near the base of the 3rd antennal segment (figure 51) ❭ 45
44.b. Arista thick, placed at the tip or dorsally near the tip (figure 52) ❭ *Pelecocera* Meigen (p. 164)

45.a. Abdomen: petiolate, constricted at tergite 2, broadened on tergite 3 (figure 53, figure 54). Male: eyes separated on the frons ❭ 46
45.b. Abdomen: elongate with parallel sides or oval (figure 55). Males: eyes separated or meeting on the frons ❭ 47

46.a. Wing: tm strongly angular; abdomen short, 2nd tergite 1-2 times as long as wide at hind margin (figure 53); leg 3 about as long as legs 1 and 2 ❭ *Neoascia* Williston (p. 147)
46.b. Wing: tm rounded; abdomen elongate, 2nd tergite 2-4 times as long as wide at hind margin (figure 54); leg 3 much longer than legs 1 and 2 ❭ *Sphegina* Meigen (p. 209)

47.a. Eyes densely haired ❭ 48
47.b. Eyes bare or nearly so ❭ 49

48.a. Face with well-defined facial knob ❭ *Cheilosia* Meigen partim (p. 54)
48.b. Face without facial knob ❭ *Psilota* Meigen (p. 192)

49.a. Abdomen: oval, reddish, black or black with grey or reddish spots ❭ 50
49.b. Abdomen: very elongated, black with yellow spots, which may be covered in grey dusting (figure 55) ❭ *Chamaesyrphus* Mik (p. 53)

50.a. Mouth edge not elongated into a long snout ❭ 51
50.b. Mouth edge elongated into a long snout (figure 56) ❭ *Rhingia* Scopoli (p. 194)

53.

55.

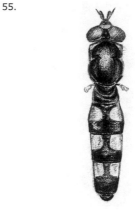

54.

56.

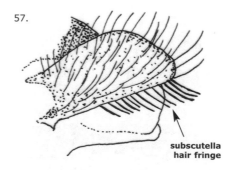

figure 53. *Neoascia tenur*, habitus of female.
figure 54. *Sphegina clunipes*, habitus of male
(Verlinden).

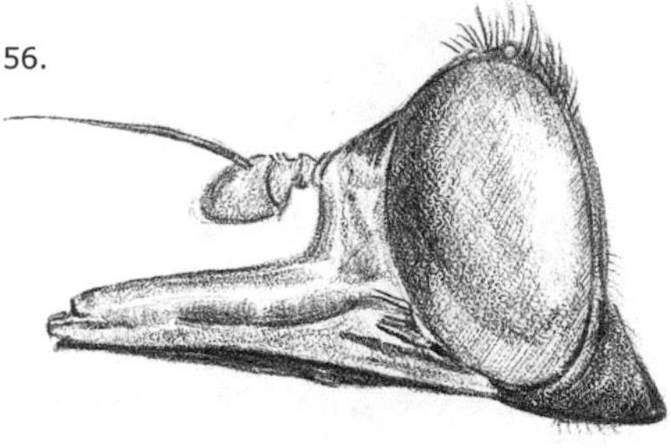

57.

subscutella
hair fringe

51.a. Abdomen: reddish-brown to orange
without clear patterning ❭ 52
51.b. Abdomen: black, metallic shiny or
black with well-marked orange or yellow pattern ❭ 53

52.a. Wing: tm ends perpendicular to
R4+5; abdomen elongated; larger
species: 10-12 mm ❭ *Hammerschmidtia*
Schummel (p. 119)
52.b. Wing: tm ends on R4+5 in a sharp
angle; abdomen rounded; smaller: 6-9
mm ❭ *Brachyopa* Meigen partim (p. 41)

figure 55. *Chamaesyrphus scaevoides*,
female.
figure 56. *Rhingia campestris*, head of male
(Verlinden).
figure 57. Scutellum with subscutellar hair
fringe present (after Thompson and
Rotheray, 1998).

58.

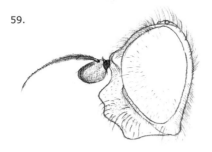

59.

figure 58. *Melanogaster aerosa*, head of female.
figure 59. *Cheilosia barbata*, head of female (Verlinden).

53.a. Wing: cell R4+5 stalked, closed well before wing margin; abdomen black, green metallic or purple **>** 54
53.b. Wing: cell R4+5 closed at the wing margin; abdomen black or black with orange spots on tergites 2 and 3 **>** *Myolepta* Newman (p. 146)

54.a. Subscutellar hair fringe present (figure 57) **>** 55
54.b. Subscutellar hair fringe absent **>** 57

55.a. Face: eye rims bordering eyes absent; facial knob often absent, as in figure 58 (only present in *Chrysosyrphus* males) **>** 56
55.b. Face: eye rims bordering eyes present (flat strip of face around inner eye margin); facial knob well-developed (figure 59)**>** *Cheilosia* Meigen partim and *Portevinia* Goffe (p.54)

56.a. Vein R4+5 with last section subequal to or longer than crossvein r-m; face with hairs laterally **>** *Chrysosyrphus* Sedman (p. 80)
56.b. Vein R4+5 with last section less than 1/2 as long as crossvein r-m; face bare laterally **>** *Lejota* Rondani (p. 126)

60.

61.

figure 60. *Chrysogaster solstitialis*, wing.
figure 61. *Orthonevra nobilis*, wing (Verlinden).

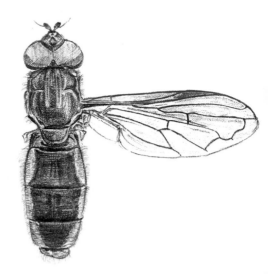

figure 62. *Riponnensia splendens*, male body and wing (Verlinden).

57.a. Tergites 2-5 completely shiny **>** 58
57.b. Tergites 2-5 dull in the middle, shiny at the margins, creating a large, median dull area on the abdomen **>** 59

58.a. Thorax and abdomen black. Males: eyes meet on frons **>** *Melanogaster* Rondani partim (p. 134)
58.b. Thorax and abdomen greenish, for one species partly purple to reddish. Males: eyes separated on frons **>** *Lejogaster* Rondani (p. 125)

59.a. Sternite 1 dull; tergite 2: dull; tergites velvety black or black with purple or green sheen; wing: tm ends perpendicular to R4+5 (and is nearly straight) or it ends in a sharp angle towards the wing tip (figure 60). Male: facial knob more or less developed **>** 60
59.b. Sternite 1 shiny; tergite 2: shiny on middle; tergites with green to bronze colour; wing: tm ends perpendicular to R4+5 (in this case, middle part of tm bent inwards) or it ends towards the wing base and angular (figure 61). Male: facial knob absent **>** 61

60.a. Antennae black; thorax and abdomen black **>** *Melanogaster* Rondani partim (p. 134)
60.b. Antennae red; thorax and abdomen black with purple or greenish sheen **>** *Chrysogaster* Meigen (p. 80)

61.a. Wing: tm ends perpendicular to R4+5 and middle part of tm bent inwards, creating an S-shaped vein (figure 62) **>** *Riponnensia* Maibach, Goeldlin & Spheight (p. 196)
61.b. Wing: tm ends towards the wing base on R4+5, the last part recessive, creating an angular vein (figure 61) **>** *Orthonevra* Macquart (p. 151)

ANASIMYIA

Introduction

Anasimyia frequent marshy areas and the banks of waterways, where they fly around the vegetation or just above the water. They visit the many flowers along the banks, such as *Mentha*, *Lysimachia* and *Iris*. *Anasimyia lineata* is the most common species and is frequently encountered on the water's edge. *A. lunulata* is the rarest, and is found in similar habitats to *Parhelophilus consimilis*. Both are

64.

curved

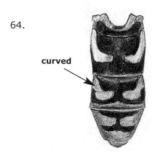

65.

triangular

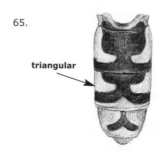

66.

hooked

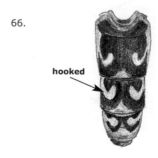

pointed

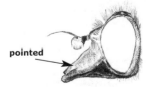

figure 63. *Anasimyia lineata*, head of male (Verlinden).

figure 64. *Anasimyia interpuncta*, male abdomen.
figure 65. *Anasimyia lunulata*, male abdomen.
figure 66. *Anasimyia transfuga*, male abdomen (Verlinden).

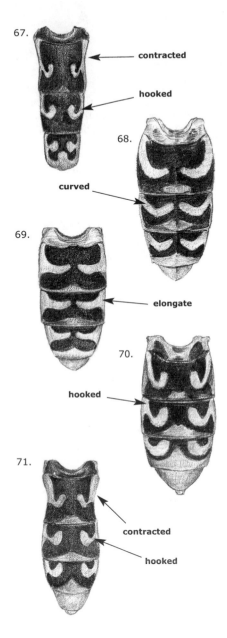

67.

contracted

hooked

68.

curved

69.

elongate

70.

hooked

71.

hooked

contracted

hooked

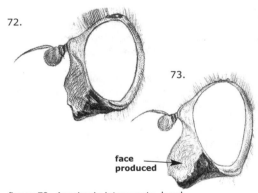

72.

73.

face
produced

figure 72. *Anasimyia interpuncta*, head
of male
figure 73. *Anasimyia lunulata*, head of male
(Verlinden).

figure 67. *Anasimyia contracta*, male
abdomen.
figure 68. *Anasymia interpuncta*, female
abdomen.
figure 69. *Anasimyia lunulata*, female
abdomen.
figure 70. *Anasimyia transfuga*, female
abdomen.
figure 71. *Anasimyia contracta*, female
abdomen (Verlinden).

peatland species, a habitat that is becoming
increasingly rare in Northwest Europe. The
long-tailed larvae live in pools and ponds
where decaying vegetation, particularly
from *Typha*, accumulates (Rotheray, 1993).

Recognition

Small hoverflies with longitudinal pale
stripes of dust on a black thorax and pale
grey, yellow or orange markings on the
abdomen. Where *Helophilus* and *Parhelophilus* appear yellowish, *Anasimyia* species
appear much paler, almost grey on black.
Face without bare stripe, in contrast to
Helophilus.

Key

1.a. Mouth edge not forming a pointed
snout (figure 72, figure 73); wing: stigma
represented by a darkened cross vein ❭ 2
1.b. Mouth edge strongly protrudes forwards like a pointed snout (figure 63);
wing: stigma represented by a dark patch
between the veins. 8-9 mm. Holarctic distribution ❭ *Anasimyia lineata* Fabricius
Jizz: face long pointed, males greyish, females yellowish.

2.a. Tergites 2-4: pale triangular markings pointing towards centre, if markings somewhat hooked their inner part not broadened (figure 64, figure 65, figure 68, figure 69) ❯ 3
2.b. Tergites 2-4: pale markings pointing forwards; markings more or less hook-shaped and inner part broadened like a droplet, occasionally markings reduced to spots; (figure 66, figure 67, figure 70, figure 71) ❯ 4

3.a. Face reaches as far forwards as antennal knob in side view. Male: tergite 4: black pattern in the form of an H; tergite 3: pale markings strongly hooked on posterior edge so that black extends forwards along lateral margin of tergite (figure 64). Female: tergites 3 and 4 with slightly hooked margins (figure 68), tergite 5 whitish-haired. 8.5-11 mm. Northern and Central Europe. ❯ *Anasimyia interpuncta* Harris
3.b. Face extends beyond antennal knob in side view. Male: tergite 4: black pattern in the form of a V or Y, front margin often dark tergite 3: pale markings straight along posterior edge so black does not extend forwards along lateral angle of tergite (figure 65). Female: tergites 3 and 4 with markings which are straight (transverse) along their posterior edge (figure 69) tergite 4 partially black-haired. 7-10 mm. Holarctic distribution. ❯ *Anasimyia lunulata* Meigen
Jizz: mouth edge extended, but not pointed (see *A. lineata*); whitish spots on abdomen.

4.a. Male: sternites 2 and 3: grey dusting with an undusted shiny stripe in the middle. Female: tergite 2: distance between front edge of pale markings to front margin of tergite larger than distance between pale markings (figure 71). 7-10.5 mm. Northern and Western Europe. ❯ *Anasimyia contracta* Claussen & Torp
Jizz: narrow species, tergite 2 strongly contracted in the middle.
4.b. Male: sternites 2 and 3 completely grey dusted. Female: tergite 2: distance between front edge of pale markings to front margin of tergite shorter than distance between pale markings (figure 70). 9-11 mm. Europe except Mediterranean, east to Western Siberia. ❯ *Anasimyia transfuga* Linnaeus
Jizz: slightly broader than *A. contracta*, tergite 2 less contracted in the middle.

ARCTOPHILA

Introduction

Arctophila frequent wet areas where water flows or seeps down from hills. They visit flowers such as *Knautia* along forest edges and in small meadows, or are seen basking in sunlight. The larvae are undescribed, but are probably aquatic/subaquatic and live among organic debris in semi-liquid mud close to streams and springs (Speight, 2003), as a female *A. fulva* was seen ovipositing in deep, water-filled hoof prints along a shaded muddy path by a stream (Stubbs and Falk, 1983).

Recognition

Large bumblebee mimics, with long hairs on the abdomen and thorax, and a crawling behaviour while visiting flowers. The wing contains a characteristic wedge-shaped dark spot in the middle. The arista is feathered with long hairs. The face is extended downwards to form a conical snout. *Arctophila* may be confused with *Pocota* and *Criorhina*, which have a bare arista or with *Volucella*, in which cell R1 is closed.

Key

1.a. Thorax and abdomen completely covered with pale, brown to yellow hairs, without bands of black hairs; legs black. 13-16 mm. Northwest and Central Europe. ❯ *Arctophila superbiens* Müller (= *Arctophila mussitans* Fabricius)
1.b. Thorax and abdomen with bands of black hair (figure 74); femora black, tibia and tarsae brown. 16-20 mm. Europe. ❯ *Arctophila bombiformis* Fallén

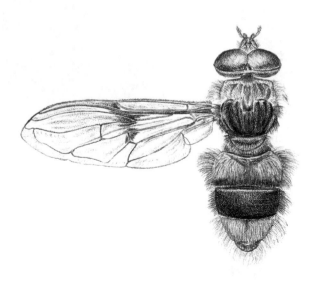

figure 74. *Arctophila bombiformis*, habitus of male (Verlinden).

BACCHA

Introduction

Baccha are exceptionally slender hover-flies, nicknamed 'flying pins', which prefer shady parts of forests. Their larvae are carnivorous and feed on aphids and other homopterans.

Recognition

The long, slender body and long wings are unique (figure 75). The only genus resembling *Baccha* is *Sphegina*, which are also found in similar, shady habitats. *Baccha* have a slender femur 3, unlike the thickened femur 3 of *Sphegina*.

Key

1. Thoracic dorsum shiny; wing hyaline to darkened, with brownish cross veins. Male: frons dusted or shiny. Western part of Palaearctic, Nearctic. ❯ *Baccha elongata* Fabricius
Note: some authors distinguish *Baccha obscuripennis* Meigen by the lack of dust on the frons of the male. Male *B. elongata* typically have a dusted frons.

figure 75. *Baccha elongata*, habitus of female (Verlinden).

figure 76. *Blera fallax*, habitus of male (Verlinden).

BLERA

Introduction

Blera are boreomontane species found in coniferous forests. The larvae and pupae are found in rotten heartwood in trunk cavities in *Pinus sylvestris* stumps (Rotheray and Stuke, 1998). *B. fallax* larvae have been found in a water-filled tree hole by Dusek and Laska (1961).

Key

1.a. Abdomen: first tergites black, last tergites red (figure 76): male tergites 1 and 2 black, female tergites 1-3 black; thorax black; wing clear. 10-12 mm. Boreomontane species in the Paloearctic. ❯ *Blera fallax* Linnaeus
1.b. Abdomen: completely black. 10-12 mm, Sweden, Siberia. ❯ *Blera eoa* Stackelberg

BRACHYOPA

Introduction

Brachyopa appear more like Muscidae or Scatophagidae than typical Syrphidae. The genus lacks any clear signs of mimicry and is also quite unlike the blackish *Cheilosia* or *Pipiza*. *Brachyopa* can be found on bark (the typical place to find species such as *B. insensilis or B. testacea*) or on flowers of Apiaceae and *Crataegus* (the typical place to find *B. testacea* or *B. vittata*). The males may be found hovering in front of injured trees. They hover like Muscidae, with an unsteady up and down motion. Females may be encountered near tree injury sites and saps runs.
The larvae of *Brachyopa* live in and from the sap runs of deciduous and coniferous trees (Rotheray, 1993), and can be found by inspecting these areas. According to experienced searchers, sap runs under bark are good places to look for the larvae. Rotheray (1996) provided a key to the larvae.

Recognition

Brachyopa are recognised by its brownish body and projecting mouth edge. They are similar to *Hammerschmidtia*, but smaller. The venation of the wings is typical for Syrphidae, which helps to distinguish it from other Diptera. Thompson (1980), Thompson and Torp (1982), Van der Goot (1986), Van Steenis (1998) and Dozckal and Dziock (2004) published (partial) keys and descriptions for this genus.

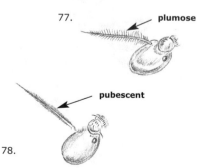

figure 77. *Brachyopa testacea*, antenna.
figure 78. *Brachyopa pilosa*, antenna (Verlinden).

Key

1.a. Arista plumose, with hairs many times longer than arista width (figure 77); mesonotum extensively pale, reddish-brown to orange **>** 2

1.b. Arista bare or pubescent, with hairs not more than twice as long as aristal width (figure 78) **>** 4

2.a Katepisternum bare; smaller. 6-7 mm: head length equal to or slightly less than its height (figure 79); anepisternum with dorsoposterior pile hair-like, usually yellowish **>** 3

2.b. Katepisternum pilose dorsally; larger, 8-9 mm; head length distinctly greater than its height (figure 80); anepisternum with black bristles or bristle-like pile dorsoposteriorly. Northern and central parts of the Palaearctic. **> Brachyopa vittata** Zetterstedt

Jizz: large *Brachyopa*, with relatively long snout and wings.

3.a. Tergite 2: dark median line present (figure 81). Males. Genitalia: sternite 9 wrinkled on its surface, in direct comparison appears wider and shorter, surstylus without bristles at tip (figure 83). 6-7 mm. Northern and central parts of the Palaearctic. **> Brachyopa testacea** (Fallén)

3.b. Tergite 2: dark median line absent (figure 82). Males. Genitalia: sternite 9 not wrinkled but flat, in direct comparison appears narrower and longer, surstylus with bristles at tip (figure 84). 6-7 mm. Northern Europe. **> Brachyopa obscura** Thompson and Torp

4.a. Thoracic dorsum extensively pale, reddish-brown to orange, laterally and anterior to scutellum **>** 5

4.b. Thoracic dorsum entirely dark, bluish-grey to black, except postalar callus usually paler **>** 6

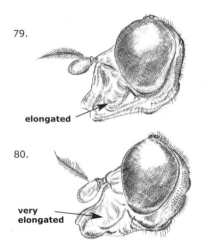

79.

elongated

80.

very elongated

figure 79. *Brachyopa testacea*, head of male
figure 80. *Brachyopa vittata*, head of male
(Verlinden).

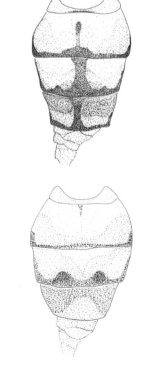

81.

82.

figure 81. *Brachyopa testacea*, abdomen
figure 82. *Brachyopa obscura*, abdomen
(after Van Steenis, 1998).

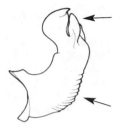

figure 83. *Brachyopa testacea*, male genitalia.
figure 84. *Brachyopa obscura*, male genitalia
(after Thompson and Torp, 1982).

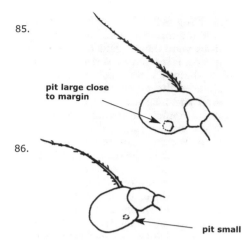

figure 85. *Brachyopa panzeri*, antenna.
figure 86. *Brachyopa dorsata*, antenna (after
Thompson, 1980).

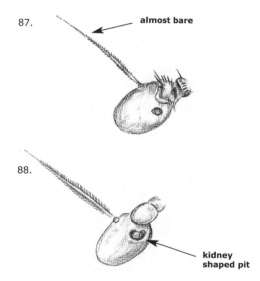

figure 87. *Brachyopa bicolor*, antenna.
figure 88. *Brachyopa scutellaris*, antenna
(Verlinden).

5.a. Antennal segment 3 with large sensory pit, with sensory pit closer to ventral margin than its diameter; male genitalia with lingular arm slender, longer than ventrolateral arm of superior lobe. 6-8 mm. Central Europe. ❯ *Brachyopa panzeri* Goffe
Note: confused with *dorsata* until recently.

5.b. Antennal segment 3 with small sensory pit, with sensory pit separated from the ventral margin by a distance greater than its diameter: male genitalia with lingular arm broad, equal to ventrolateral arm. 6-8 mm. Central and Northern Europe, northern parts of Asia to the Pacific. ❯ *Brachyopa dorsata* Zetterstedt

6.a. Arista bare or virtually so, with hairs less than aristal width (figure 87) ❯ 7
6.b. Arista pubescent, with some hairs as long as basal aristal width (figure 88) ❯ 14

7.a. Wing hyaline, without dark spots; thorax except scutellum entirely dark, from bluish-grey to black; male: line of contact of eyes short, about as long as ocellar triangle; female front entirely pale pilose ❯ 8

7.b. Wing with distinct black spots on anterior crossvein (r–m) and end of spurious vein; thorax dark only on pectus (pectus is the ventral surface of the thorax between the legs and, as such, includes the ventral 2/3 of katepisternum and meron and all of katatergite) and thoracic dorsum; postpronotal lobes, anepisternum, dorsal katepisternum, anterior anepimeron, barrette (dorsal raised edge of meron), pleurotergum all pale reddish-brown to orange; male eye contiguity long, longer than ocellar triangle; female front with a few black hairs; antennal segment 3 with medium-sized sensory pit. 6–8 mm. Central Europe. ❯ *Brachyopa maculipennis* Thompson
Jizz: wing with dark spots.

8.a. Abdomen, middle and hind coxae dark, brownish-black to black; scutellum orange, short adpressed black pilose. Male: eyes distinctly separated by much more than aristal width; proepimeron pilose; 3rd antennal segment without a distinct sensory pit ❯ 9
8.b. Abdomen, middle and hind coxae pale, reddish-brown to orange; male eyes meeting or approximate, in the latter case separated by less than aristal width; proepimeron bare ❯ 10

9.a. Abdominal tergites with hind margins pollinose and pale, yellow to white; scutellum black on basal 2/3, orange apically, entirely long erect pale yellow pilose. 6–8.5 mm. Siberia ❯ *Brachyopa ornamentosa* Violovitsh
9.b. Abdominal tergites shiny, uniformly dark; scutellum orange, short appressed black pilose. 6–8 mm. Northern part of the Palacarctic. ❯ *Brachyopa cinerea* Wahlberg
Jizz: abdominal tergites shiny, uniformly dark.

10.a. Antennal segment 3 without a sensory pit (figure 89) or sensory pit minute, its diameter less than the aristal width; notopleuron extensively pale pilose ❯ 11
10.b. Antennal segment 3 with a small sensory pit (figure 87); notopleuron black pilose. 6–9 mm. Northern and central parts of the Palaearctic. ❯ *Brachyopa bicolor* (Fallen)

11.a. Scutellum entirely covered in microtrichia; femur 3 thick: length:width ratio in males smaller than 4.5, in females smaller than 5. Female: tergite 2 entirely pale haired ❯ 12
11.b. Scutellum largely bare of microtrichia, except for anterior margin; femur 3 slender: length:width ratio in males larger than 5, in females larger than 5.5. Female: tergite 2 with extensive black hairs. 6–9 mm. Western and Central Europe, southeast to Tadjikistan. ❯ *Brachyopa insensilis* Collin

12.a. Hypostomal bridge (mid ventral plate of the head) blackish; thoracic dorsum either with undusted triangular spots at the suture (*B. silviae*) or bare spots absent (*B. grunewaldensis*); tibia 1 and 3 and dorsal surface of tars 1, 2 and 3 entirely pale haired with at most a few black hairs mixed in; scutellum black haired with few pale hairs anteriorly ❯ 13
12.b. Hypostomal bridge yellow; thoracic dorsum with one pair of undusted roundish spots at the transverse suture; tibia 1 and 3 and tars 1, 2 and 3 black haired; scutellum pale haired. 7 mm. Central and Southeastern Europe ❯ *Brachyopa bimaculosa* Doczkal and Dziock 2004

13.a. Thoracic dorsum with a pair of undusted triangular spots at the transverse suture; face more protruding; head: bare spots posterior to the posterior ocelli lack; postalar calli entirely microtrichose. Male: apex of tibia 3 ventrally with a narrow, sharp edge acros full width. 8mm. Central Europe. ❯ *Brachyopa silviae* Doczkal and Dziock 2004
13.b. Thoracic dorsum without bare spots; face less protruding; head: a bare spot is present posterolateral to posterior ocelli; postalar calli with a spot bare of microtrichia. Male: apex of tibia 3 ventrally with a narrow sharp edge restricted to the posterior threequarters of the apical margin. 8mm. Central Europe. ❯ *Brachyopa grunewaldensis* Kassebeer 2000

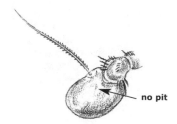

figure 89. *Brachyopa insensilis*, antenna (Verlinden).

no pit

14.a. Antennal segment 3 with a smaller, rounded sensory pit (figure 78) -> 15
14.b. Antennal segment 3 with a large, kidney-shaped sensory pit (figure 88); tergite 2 posteriolaterally black pilose. 6-8 mm. Western and Central Europe. ❯ *Brachyopa scutellaris* (Robineau-Desvoidy)
Jizz: humerus brown, hardly dusted.

15.a. Antennal segment 3 with large sensory pit, with sensory pit closer to ventral margin than its diameter (figure 90); tergite 2 black pilose posteriolaterally. Central and Eastern Europe ❯ *Brachyopa plena* Collin
15.b. Antennal segment 3 with small sensory pit, with sensory pit further removed from ventral margin, separated by its diameter (figure 78); tergite 2 often entirely pale pilose, sometimes intermixed with a few black. 6-8 mm. Europe except Mediterranean. ❯ *Brachyopa pilosa* Collin

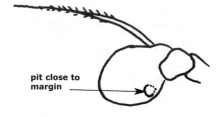

pit close to margin

figure 90. *Brachyopa plena*, antenna (after Collin).

BRACHYPALPOIDES

Introduction

Brachypalpoides inhabit damp deciduous forests, where they walk on leaves and bark, occasionally visiting flowers. They are restless sawfly-mimics, closely allied to *Xylota* and *Chalcosyrphus*. It can be confused with *Chalcosyrphus piger* which has the abdomen red from tergite 2 on, the tip is red.

Key

1. Only one European species. Legs black; thorax black; abdomen black with tergites 2 and 3 mainly red, tip of abdomen black. 11-13.5 mm. Europe, east into Asia-minor. ❯ *Brachypalpoides lentus* Meigen

BRACHYPALPUS

Introduction

Brachypalpus are found on tree bark or visiting flowers. *B. laphriformis* and *B. valgus* visit flowering trees and shrubs, such as *Crataegus* and *Sorbus* (De Buck, 1990; personal observations), while *B. chrysitis* is also found on flowering herbs, such as *Apiaceae*, *Petasitis albus* and *Ranunculus*. The larvae live in rot holes in deciduous trees (Rotheray, 1993). *B. chrysitis* occurs in *Abies/Picea* forests and its undescribed larvae may live in coniferous trees.

Recognition

Brachypalpus are bee-like in appearance, as they are brownish and haired. However, their behaviour is completely different from that of bees, as they run on leaves and bark. Their hind femur is enlarged. When viewed from the front the head is triangular (more distinct in males than females), quite unlike *Criorhina* and *Xylota*. The key is based on Van der Goot (1981), Verlinden (1991) and Speight (1999).

Key

1.a. Males (eyes approaching or meeting above antennae) **>** 2
1.b. Females (eyes not meeting above antennae) **>** 4

2.a. Hind tibiae abruptly angled at about 1/3 from its distal end and with a triangular flange projecting strongly from its ventral surface at about 1/3 from the base of the tibia (figure 91); hairs clustered around middle of posterolateral surface of hind tibiae longer than the maximum width of a hind tibia **>** 3

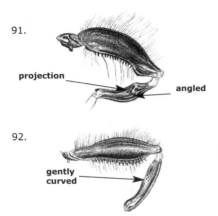

91.

projection

angled

92.

gently curved

figure 91. *Brachypalpus valgus*, femur and tibia 3.
figure 92. *Brachypalpus laphriformis*, femur and tibia 3 (Verlinden).

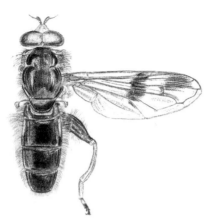

figure 93. *Brachypalpus laphriformis*, habitus of male (Verlinden).

2.b. Hind tibiae gently curved in apical 1/3: no hairs on the hind tibiae as long as the maximum width of a hind tibia (figure 92); hairs on general body surface mostly sandy brown; tergites 2-4 entirely undusted, brightly shining (figure 93). 10-13 mm. Central Europe **>** *Brachypalpus laphriformis* Fallén (= *Brachypalpus meigeni* Schiner)
Jizz: elongate, brownish species with a triangular head in front view, mouth edge not pronounced.

3.a. Hairs on general body surface rufous; abdominal tergites 2 and 3 undusted, brightly shining, but each with a pair of dull narrow black bars of dusting; tibia 3 strongly angled (figure 94). 13-15 mm. Northern and Central Europe **>** *Brachypalpus chrysites* Egger
Jizz: abdomen with long, but not very dense yellow hairs.

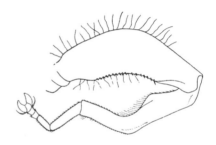

figure 94. *Brachypalpus chrysites*, femur and tibia 3 (after Stackelberg).

3.b. Hairs on general body surface very pale brownish-yellow; abdominal tergites 2 and 3 thinly dusted, dark grey over most of surface and rather dull, but each with a pair of transverse black bars that are entirely undusted and brightly shining; tibia 3 less angled (figure 91). 11-14 mm. Central Europe **>** *Brachypalpus valgus* Panzer
Jizz: eyes do not meet, tip of abdomen black-haired.

4.a. Hair covering on abdomen reddish-yellow; hair on abdominal tergite 3 semi-erect, on disc more than 1.5 times as long as the maximum depth of a hind tibia. 13-15 mm. Northern and Central Europe **>** *Brachypalpus chrysites* Egger
Jizz: abdomen with long, but not very dense yellow hairs.

4.b. Hair covering on abdomen whitish-yellow/brownish and black; hair on abdominal tergite 3 erect, on disc no longer than maximum depth of a hind tibia (hair at lateral margins longer) ❭ 5

figure 95. *Brachypalpus laphriformis*, habitus of female (Verlinden).

5.a. Frons undusted, shiny across most of its width, only dusted narrowly against the eyes; notopleural area and indented line of the transverse suture on the thoracic dorsum undusted, shiny; hind coxae black; hind tars with 2nd segment about twice as long as its maximum width. 11-15 mm. Central Europe ❭ **Brachypalpus valgus** Panzer

5.b. Frons mostly covered in dusting, the 2 large dust spots almost meeting in the mid-line; notopleural area and indented line of the transverse suture on the thoracic dorsum heavily dusted grey and dull; ventral surface (at least) of the hind coxae yellow; hind tars with 2nd segment distinctly more than twice as long as its maximum width. 10-13 mm (figure 94). Central Europe ❭ **Brachypalpus laphriformis** Fallén

Jizz: elongate, brownish species with a triangular head in front view, mouth edge not pronounced.

CALIPROBOLA

Introduction

Caliprobola speciosa is a species of ancient woodlands. Its larvae live in decaying beech (*Fagus*) wood. There is nothing else like this golden green, shiny and large hoverfly in Northwest Europe.

Key

1. Face yellow, mouth edge black; thorax, scutellum and abdomen black with strong metallic golden-green shine; legs yellowish, basal part of femora black; wing yellowish infuscated at front margin, with dark patch in tip (figure 96). 13-15 mm. Central part of Palaearctic, in Europe north to Denmark and south to Northern Italy ❭ *Caliprobola speciosa* Rossi

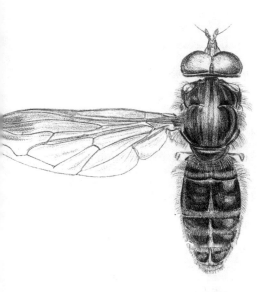

figure 96. *Caliprobola speciosa*, habitus of male (Verlinden).

CALLICERA

Introduction

Callicera inhabit forest and are thought to live in the canopy. Adults may be found hovering 2-3 metres above the ground, walking or sitting on bark and leaves. They are rarely seen. The larvae of *Callicera* inhabit rotten hardwood. Females deposit their eggs in rot holes.

Recognition

Callicera are large, conspicuous flies, with long antennae and a black or metallic shiny body, which resemble bees. Their arista is implanted on top of the third antennal segment, a very unusual position among Syrphidae. Speight (1991) published a key for this genus, that is largely followed here.

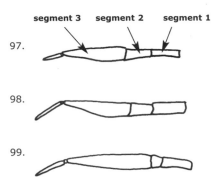

figure 97. Callicera aenea, antenna.
figure 98. Callicera spinolae, antenna.
figure 99. Callicera fagesii, antenna (Speight, 1991).

Key

1. Antennal segment 2 not less than 3/4 the length of antennal segment 1 (segment 2 may be slightly longer than segment 1); antennal segment 3 about as long as segments 1 + 2 together (figure 97) **>** 2
1.b. Antennal segment 2 no more than 1/2 as long as antennal segment 1; antennal segment 3 twice as long as segments 1 + 2 together (figure 99) **>** 4

2.a. Male: frons with long hairs along eyes above antennae (as long as the hairs on adjacent parts of the eyes); tergite 3 with a complete, dull black, transverse band close to its posterior margin. Female: femora entirely yellow-orange; tergite 3 with a dull black, transverse band close to its posterior margin. 11-13 mm. From Great Britain to the south of Russia. **>** *Callicera spinolae* Rondani
2.b. Male: frons without hairs above antennae (other than microscopic pile); tergite 3 shiny all over or with a pair of matt black marks. Female: legs with femora extensively black; tergite 3 shiny all over, brassy, or with a pair of narrow, transverse, matt black marks within the posterior half of the tergite (figure 100, figure 101) **>** 3

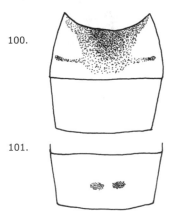

figure 100. *Callicera aenea*, tergite 2 and 3 of male.
figure 101. *Callicera aenea*, tergite 3 of male (Speight, 1991).

3.a. Thoracic dorsum thinly but distinctly dusted greyish almost over entire surface, back to scutellum (with or without 2 or 4 distinct longitudinal dust-stripes within the general dusting); hairs on scutellar disc 2/3 or more the length of the scutellum and hairs on posterior margin of scutellum as long as the scutellum; pleura entirely pale-haired; legs usually entirely pale-haired; all trochanters pale-haired; hairs on posterolateral surface of fore and mid tibiae including some up to as long as the maximum width of a tibia in dorsal view; all tarsal segments as yellow as their metatarsus or tarsal segments 3-5 only vaguely greyish, much paler than the blackened basal parts of the femora. 11-13 mm. Northern and central parts of the Palaearctic. **>** *Callicera aenea* (Fabricius)

3.b. Thoracic dorsum brightly shiny, except for 2 longitudinal grey stripes of dusting medially, which stop abruptly between the wing bases (and so do not reach the scutellum); hairs on scutellum nowhere as long as the scutellum and on scutellar disc no more than 1/2 as long as the scutellum; pleura usually black-haired ventrally; at least fore and hind trochanters partly black-haired and at least fore femora extensively black-haired posterolaterally; hairs on posterolateral surface of fore and mid tibiae all shorter than maximum width of a tibia in dorsal view; tarsal segments 3-5 of all legs almost black, as dark as the blackened basal parts of the femora. 11-13 mm. **>** *Callicera aurata* Rossi

4.a. Hind femora smoothly curved and flat, ventrally (figure 102b,d); legs usually entirely orange, except for last 2 tarsal segments, which are black on all legs (the femora may be narrowly black at the base and the hind femora may be black as far as the basal 1/3); thoracic dorsum undusted, shiny; wings with extensive areas of membrane bare of microtrichia on basal cells and anal cell; tergite 3 with a transverse, dull black band, interrupted at the middle, close to the posterior margin of the tergite (figure 103). 10-12 mm. Western and Central Europe. **>** *Callicera rufa* Schummel

4.b. Hind femora angled ventrally, ventral surface shallowly concave in apical 1/2 of length (figure 102, a, c); legs with at least all femora mostly black and tarsal segments 3-5 of all legs black; thoracic dorsum with longitudinal stripes of grey dusting medially; wings almost entirely covered in microtrichia (often a narrow strip bare along anterior margin of anal cell and a small patch bare along midline of 2nd basal cell) **>** 5

5.a. Scutellar disc with many hairs no longer than antennal segment 1; hairs on thorax and abdomen orange to whitish-yellow, straight; abdominal tergites almost without black hairs; all tibiae orange; genae beneath eyes proportionally narrower. 10-12 mm. Southern Europe **>** *Callicera macquarti* Rondani

5.b. Hairs on scutellar disc all longer than antennal segment 1; hairs on thorax and abdomen yellow-brown to grey-brown, somewhat wavy; at least tergite 4 often extensively black-haired (tergites 3 and 4 and apical abdominal sternites may be extensively black-haired); tibiae may be all orange, but at least hind tibiae often brownish or partly black; genae beneath eyes proportionally wider. 10-12 mm. Central Europe **>** *Callicera fagesii* Guerin-Meneville

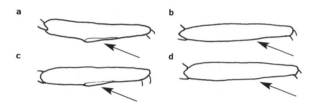

figure 102. *Callicera, femur 3*: (a) C. fagesii; (b) C. rufa, male; (c) C. macquartii; (d) C. rufa, female (Speight, 1991).

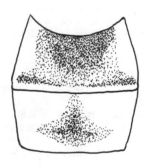

figure 103. *Callicera macquartii*, tergite 2 and 3 of male (Speight, 1991).

CERIANA AND SPHIXIMORPHA

Introduction

Ceriana and *Sphiximorpha* inhabit mature deciduous forests with old growth. They are also regularly found in parks containing senescent trees (*S. subsessilis* in the

Rosenstein park in Stuttgart (Schmidt, 1996); *C. vespiformis* in the Jardin Botanico in Madrid (personal observation)). Their larvae appear to live in sap runs and moist tree holes in deciduous trees, notably *Populus* and *Ulmus* and probably also *Betula* (Speight, 2003).

Recognition

Ceriana and *Sphiximorpha* are convincing wasp mimics, with their long bodies, yellow and black patterning and long antennae. Arista on top of third antennal segment.

Key

1.a. Antennae implanted on a long prominence on the frons, that is as long as the 1st antennal segment; wing (figure 104): vein R4+5 sharply curved into the underlying cell ❯ 2
1.b. Antennae implanted on the frons, prominence absent (figure 105); wing: vein R4+5 almost straight. 12-14 mm. Central Europe, east to the Ural ❯ *Sphiximorpha subsessilis* Illiger

2.a. Scutellum yellow with black tip; larger species: 11-14 mm (figure 106). Palaearctic ❯ ***Ceriana conopsoides*** Linnaeus
2.b. Scutellum entirely yellow; smaller species: 8-10 mm. Southern parts of Europe, North Africa; some specimens collected in The Netherlands ❯ ***Ceriana vespiformis*** Latreille

CHALCOSYRPHUS

Introduction

Most *Chalcosyrphus* look like sawflies (Symphyta) and are found walking on leaves. *C. eunotus* is an exception, being more bee-like in appearance. *Chalcosyrphus* seems to prefer damp environments. *C. nemorum* is abundant in damp, deciduous forests in The Netherlands, e.g.

Alnus forest on peatland. *C. eunotus* is typically found along the edges of streams in deciduous forests (Schmidt, 1996), where they settle on logs and stones near the water.
The larvae live in sap runs under bark and in rot holes. *C. nemorum* larvae are found behind the bark of water-sodden deciduous timber, stumps and in damp rot holes in deciduous trees, such as *Betula*, *Fagus*, *Populus*, *Quercus*, *Salix* and *Ulmus* (Speight, 2003). *C. piger* larvae are found under the bark of *Pinus*.

Recognition

Chalcosyrphus resemble *Xylota* and there are only subtle differences between the adults of these genera (see key to genera p. 32). *Chalcosyrphus* larvae bear hooks on their heads, which are lacking in *Xylota*. There are three morphological groups of *Chalcosyrphus* in Northwest Europe. First, a group with black legs (*C. nigripes*, *C. piger*), second, those with mainly red hind femora and black abdomens (*C. femorata*, *C. valgus*, *C. rufipes*), and third, a group of short species whose abdominal tergites are much wider than they are long (*C. nemorum*, *C. eunotus*, *C. jacobsoni*). The key is based on Van der Goot (1981), Verlinden (1991) and Speight (1999).

Key

1.a. Legs black ❯ 2
1.b. Legs with pale markings ❯ 3

2.a. Abdominal tergites 2 and 3 black with yellow spots or shiny black patches where spots should be; femur 3 strongly swollen (figure 107). 10 mm. Arctic parts of the Palaearctic ❯ ***Chalcosyrphus nigripes*** Zetterstedt
2.b. Abdominal tergites 3-5 entirely orange-red (typical specimens) or uniformly black (currently only known from Corsica); femur 3 less swollen (figure 108). 11-13 mm. Northern and Central Europe, east to Pacific coast, Nearctic. ❯ ***Chalcosyrphus piger*** Fabricius
Note: If legs black and tergite 3 red, but thoracic dorsum finely punctuate and tergite 5 black › *Brachypalpoides lentus (p. 32)*.

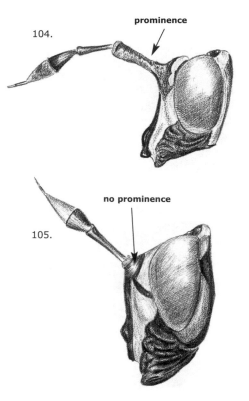

104.

prominence

105.

no prominence

figure 104. *Ceriana conopsoides*, head of male.
figure 105. *Sphiximorpha subsessilis*, head of male (Verlinden).

figure 106. *Ceriana conopsoides*, habitus of male (Verlinden).

107.

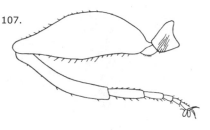

108.

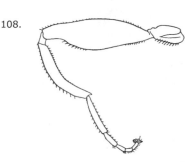

figure 107. *Chalcosyrphus nigripes*, leg 3 of male.
figure 108. *Chalcosyrphus piger*, leg 3 of male (after Stackelberg).

3.a. Abdominal tergite 2 as long as wide or longer than wide (figure 109); femur 3 largely reddish to orange **›** 4
3.b. Abdominal tergite 2 wider than long (figure 110); femur 3 largely black **›** 7

4.a. Distal ends of hind femora black (figure 111); hind tibiae and tarsi entirely black **›** 5
4.b. Femur 3 entirely orange; hind tibiae and tarsi brownish distally. 10-12 mm. Continental part of Central Europe **›** *Chalcosyrphus pannonicus* Oldenberg

5.a. Halter knob grey-brown to dark brown; apex of the hind tibia flat or with a short keel postero-ventrally, terminating in a minute spike, which is distinctly shorter than 1/4 the apical width of the hind tibia (figure 111). Male: no hairs on the anterior pair of tibiae as long as the width of the tibia in dorsal view **›** 6
5.b. Halter knob pale yellow; apex of the hind tibia extended ventrally into a large triangular flange (reminiscent of a mortar-trowel blade), which is as long as 1/2 the apical width of the tibia (figure

figure 109. *Chalcosyrphus valgus*, habitus of male (Verlinden).

figure 110. *Chalcosyrphus eunotus*, habitus of male (Verlinden).

112); wing: stigma dark brown. Male: hair fringe on the postero-lateral surface of the fore tibiae longer than the width of the tibia in dorsal view. 15-18 mm. Central Europe, in Asia to Pacific coast **> *Chalcosyrphus femoratus*** Linneaus (= *Chalcosyrphus curvipes* auctorum)

6.a. Arista entirely yellow-brown; stigma usually yellow-brown, but may be darker distally in old specimens; halter knob grey-brown. Male: abdominal tergite 4 nearly 1.5 times as long as abdominal tergite 3 (figure 109). Female: apex of the hind tibia with a short, but distinct, keel postero-ventrally, terminating in a minute spike, which is distinctly shorter than 1/4 the apical width of the hind tibia (figure 111). 12-14 mm. Northern and Central Europe, in Asia to Pacific coast **> *Chalcosyrphus valgus*** Gmelin (= *Chalcosyrphus femoratus* auctorum)

6.b. Arista dark brown/black apically, but yellow-brown on basal 1/2 of length: stigma dark brown: halter knob dark brown. Male: abdominal tergites 3 and 4 of almost equal length. Female: apex of hind tibia flat ventrally, without either a postero-ventral keel or spike. 12-14 mm. Continental part of Central Europe, in Asia into Mongolia **> *Chalcosyrphus rufipes*** Loew

7.a. Arista pale brownish-yellow; abdominal tergites without pale marks (patches of dense grey-dusting may be present). Male: eyes meeting for a distance shorter than 1/2 the length of the frons (figure 113) **>** 8

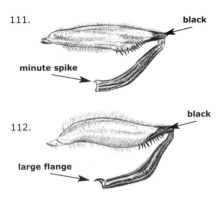

111.

black

minute spike

112.

black

large flange

figure 111. *Chalcosyrphus valgus*, femur and tibia 3.

figure 112. *Chalcosyrphus femoratus*, femur and tibia 3 (Verlinden).

7.b. Arista dark brown to black; at least abdominal tergites 2-4 with a pair of pinkish/orange-brown marks (figure 114). Male: eyes meeting for a distance longer than 1/2 the length of the frons (hair length features as in *C. jacobsoni*). 8-10 mm. Northern and Central Europe, in Asia into Japan, Nearctic **> *Chalcosyrphus nemorum*** Fabricius
Jizz: small, with short abdomen, square yellow spots and a relatively large thorax.

8.a. Body: long-haired, many of the hairs on the scutellum distinctly longer than the median length of the scutellum; hairs on the posterolateral surface of the front tibiae including some distinctly longer than the maximum width of the

figure 113. *Chalcosyrphus eunotus*, face of male, front view (Verlinden).

figure 114. *Chalcosyrphys nemorum*, habitus of male (Verlinden).

front tibia in dorsal view; hairs on posterolateral surface of hind femora as long as the maximum width of the hind femur in dorsal view; thoracic dorsum with an incomplete, transverse band of black hairs mixed in among the pale hairs at the level of the wing-bases. Male: abdominal tergite 2 with a pair of dense, more or less rectangular, grey dust spots (figure 110) in the place of the pale marks found in *C. nemorum* (these dust-marks become progressively thinner towards the lateral margins of the tergite, so that its surface may be shiny for up to 1/3 of its width). 9-11 mm. Central Europe, in Asia into Caucasus ❯ *Chalcosyrphus eunotus* Loew
8.b. Body: short-haired, hairs on scutellum at most as long as median length of the scutellum; hairs on the posterolateral surface of the front tibiae all shorter than the maximum width of a front tibia in dorsal view; hairs on posterolateral surface of hind femora distinctly shorter than the maximum width of a

hind femur in dorsal view; thoracic dorsum entirely pale-haired. Male: abdominal tergite 2 with a pair of undusted, mirror-like, brightly shining metallic patches in the place of the pale marks found in *C. nemorum*. 8-10 mm. Northern Europe, Siberia, Carpathian Mountains ❯ *Chalcosyrphus jacobsoni* Stackelberg

CHAMAESYRPHUS

Introduction

Chamaesyrphus are small, inconspicuous hoverflies. *C. lusitanicus* can be caught near *Pinus* trees in open sandy areas, and around dunes and heathers. They fly in the vegetation and visit *Calluna* flowers.

Recognition

The small size and elongated abdomen with, in many cases, pale spots are characteristic of this genus. They resemble *Platycheirus*, but their humeri are haired.

Key

1.a. Anterior anepisternum completely bare ❯ 2
1.b. Anterior anepisternum: along upper margin with erect, white hairs; lunula with shiny black upper margin, but inner parts greyish dusted; abdomen: black with reddish brown, greyish dusted spots on tergites 2-4, which may be absent. 5-6 mm. Western Europe (Portugal to Finland), European Russia. ❯ *Chamaesyrphus lusitanicus* Mik

2.a. Lunula: lower central part greyish dusted, upper margin shiny black; posterior anepisternum: white bristles present on hind upper corner; abdomen black with reddish-yellow spots, not or hardly dusted; legs yellow, at most slightly darkened in the middle (figure 115, figure 116). 4-5 mm. Northern and Central Europe, east to the Ural ❯ *Chamaesyrphus scaevoides* Fallén

figure 115. *Chemaesyrphus scaevoides*, habitus of male (Verlinden).

2.b. Lunula: completely black shiny; posterior anepisternum: white bristles absent; abdomen: yellow spots often covered with greyish dust; femora and tibiae broadly darkened in the middle. 4-5 mm. Scotland, single observations from The Netherlands and Leningrad ❭ *Chamaesyrphus caledonicus* Collin

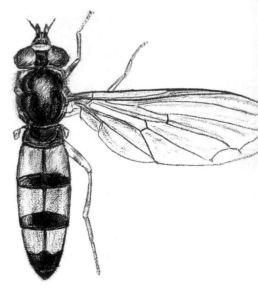

figure 116. *Chamaesyrphus scaevoides*, habitus of female (Verlinden).

CHEILOSIA

Introduction

Cheilosia is a large genus of blackish hoverflies. Adults are often found visiting flowers of Umbelliferae, Compositae and Ranunculaceae, but occur on a wide range of other flowering plants. Males may agressively defend an air space from a perch on a leaf or while hovering. Some species, like females of *C. albitarsis*, seem to be particularly attracted to the flowers of the larval food plant, in this case *Ranunculus*.

The larvae are phytophagous or fungiphagous. The majority of species is phytophageous, fungiphagous species occur in a small group related to *C. scutellata*. The food plants of many species are still unknown. Known larval food plant preferences show that polylectic, monolectic and oligolectic species occur. Polylectic species like *C. vernalis* live on a range of food plants, e.g. *Achillea*, *Sonchus*

and *Matrichia*. Other species are mono- or oligolectic and specialise on a single species or genus of plant. Specialisation even occurs within a plant species. As an example, Stuke and Claussen (2000) elucidate the larval habitats of *C. canicularis, C. himantopus* and *C. orthotricha*. The larvae of these species all live in *Petasites hybridus*, but prefer different parts. *C. canicularis* lives in the developing shoots, *C. himantopus* in the petioles of leaves and *C. ortotricha* in the flowering stems. *Petasites hybridus* is the only food plant of *C. canicularis* and *C. ortotricha*, while *C. himantopus* occurs in other species of *Petasites* as well.

Recognition

Cheilosia are blackish hoverflies, rarely with greyish spots. They are never strongly greenish or shiny purple (as *Lejogaster*, *Chrysogaster* and *Orthonevra* are) or with their abdomen dull medianly and shiny laterally (as *Melanogaster* and *Orthonevra* are); there is an eye rim along the facial border of the eyes (lacking in all other genera); and males and females have a facial knob and a mouth edge (unlike *Pipiza, Pipizella, Heringia, Neocnemadon* and females of *Lejogaster, Melanogaster* and *Chrysogaster*). There are two pitfalls. The first is that the scutellum of some females is partly yellow causing them to be mistaken for a member of the Syrphinae. In virtually all Syrphinae the face is at least partly yellow, while it is black in *Cheilosia* (with a few exceptions such as *C. pallipes*) and Syrphinae never have an eye rim. The second is that *Melangyna quadrimaculata* and *M. cingulata* may have a black scutellum and face, and may then be mistaken for a *Cheilosia*. However, they lack eye rims.

The key uses a number of species groups to order the species by morphological similarity. These groups are introduced for practical reasons and have no taxonomic significance. The key is based on Sack (1932), Nielsen (1970), Van der Goot (1981), Claussen and Speight (1988), Verlinden (1991), Bradescu (1991), Vujic and Claussen (1994a, 1994b), Barkalov and Stahls (1997), Doczkal (2000), Stuke and Claussen (2000), Zeegers (2000), Nielsen and Claussen (2001), Smit (2002), Bartsch et al. (2009b).

Key

1.a. Eyes bare (at high magnifications small hairs may be visible, these are at most as long as the width of an eye facet) **>** 2
1.b. Eyes haired **>** 4

2.a. Side of facial knob bare, at most with very short hairs (much shorter than hairs along eye rim) **>** 3
2.b. Side of facial knob with long hairs, similar to or longer than hairs on eye rim **> Angustigenis-group** (p. 56)

3.a. Legs black, at most tibia pale at the base or some tarsal segments pale **> Antiqua-group** (p. 57)
3.b. Legs partly pale, at least tibia pale at both ends **> Pagana-group** (p. 59)

4.a. Face with long hairs on the sides of the facial knob, equal to or longer than hairs on eye rim **>** 5
4.b Face without long hairs on the facial knob, at most with hairs much shorter than hairs on eye rim **>** 6

5.a. Legs black **> Variabilis-group** (p. 62)
5.b. Legs partly pale, at least tibia pale at both ends **> Illustrata-group** (p. 63)

6.a. Legs completely or predominantly black, at most middle segments of tarsae pale or the bases of the tibia pale, but the tip of the tibia black **> Impressa-group** (p. 65)
6.b. Legs partly pale, at least tibia pale at both ends **>** 7

117.

118.

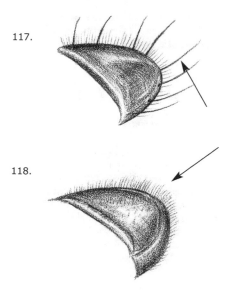

figure 117. *Cheilosia*, scutellum with bristles along margin.
figure 118. *Cheilosia*, scutellum without bristles along margin (Verlinden).

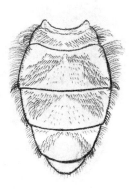

figure 119. *Cheilosia proxima*, abdomen of female (Verlinden).

7.a. Hind margin of scutellum and postalar lobes with bristles, bristles often
entirely black but sometimes partly or
entirely yellow (figure 117) **>** 8
7.b. Hind margin of scutellum without
bristles, with hairs only (figure 118),
postalar lobes with or without bristles **>**
Canicularis-group (p. 76)

8.a. Sternites dusted, females often with
elongate areas covered with longer hairs
on tergites (figure 119) **> Proxima-
group** (p. 68)
8.b. Sternites shiny, females without areas
with longer hairs on tergites **>**
Bergenstammi-group (p. 71)

Angustigenis group

1.a. Tibiae: entirely black, only pale at
very base; tars 1 brown; eyes completely
bare. 7-8 mm. Northern Europe, Siberia
> *Cheilosia angustigenis* Becker
1.b. Tibiae: broadly yellow at base, narrowly pale at tip, black in between; eyes
with very sparse, short hairs. 7-8 mm.
Palaearctic **> *Cheilosia latifrons***
Zetterstedt (= *Cheilosia intonsa* Loew)
Note: normally the eyes are haired, but this may
be overlooked, see *illustrata*-group.

121.

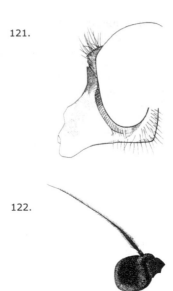

122.

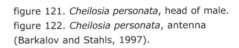

figure 121. *Cheilosia personata*, head of male.
figure 122. *Cheilosia personata*, antenna
(Barkalov and Stahls, 1997).

123.

124.

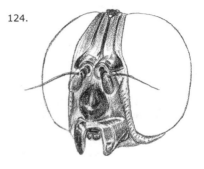

figure 123. *Cheilosia albitarsis*, tars 1.
figure 124. *Cheilosia albitarsis*, head of
female (Verlinden).

figure 120. *Portevinia maculata*, habitus of
male (Verlinden).

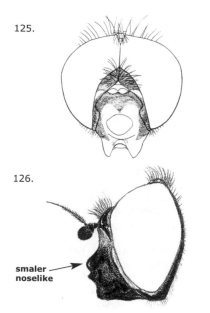

125.

126.

smaler noselike

figure 125. *Cheilosia vicina*, head of male (Barkalov and Stahls, 1997).
figure 126. *Cheilosia nigripes*, head of male (Verlinden).

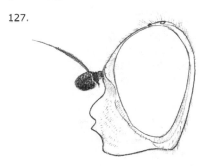

127.

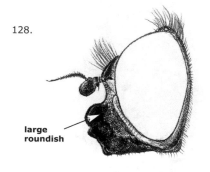

128.

large roundish

figure 127. *Cheilosia nigripes*, head of female.
figure 128. *Cheilosia antiqua*, head of male (Verlinden).

Antiqua Group

1.a. Tergites without grey spots; antennae often dark **>** 2
1.b. Tergites with grey spots (figure 120); antennae red; males: eyes meet at a single point. 8-9 mm. Northern and Central Europe **>** *Portevinia maculata* Fallén

2.a. Wing bases strongly yellowish **>** 3
2.b. Wing bases hyaline to blackish **>** 4

3.a. Males and females. Tars 1 completely black; mouth edge conically protrudes downwards (figure 121); arista with long hairs (figure 122); 8-11 mm. Mountains of Central Europe **>** *Cheilosia personata* Loew
   Jizz: from a distance like *C. variabilis*, but mouth edge protruding strongly downwards. Giant *Ch. semifaciata*.
3.b. Females. Tars 1: middle segments pale (in rare cases black) (figure 123); mouth edge not protruding downwards (figure 124); arista with length of hairs equal to diameter of arista; 7-9 mm. Palaearctic **>** *Cheilosia albitarsis* Meigen or *Cheilosia ranunculi* Doczkal
   Note: females of *albitarsis* and *ranunculi* cannot be distinguished.

4.a. Sternites 2-5 shiny, often in contrast to a dull sternite 1 **>** 5
4.b. Sternites 1-5 dull **>** 7

5.a. Posterior anepisternum at least partly shiny, with very faint dusting on parts; anterior anepisternum dusted, sometimes only weakly. Male: frons shiny **>** 6
5.b. Posterior and anterior anepisternum dusted (view from slightly above and compare with shiny notopleurae); arista short-haired. Male: frons dusted (figure 125); thoracic dorsum with short yellowish and black hairs and long black hairs. Female: scutellar margin with 2-4 bristles; thoracic dorsum with very short pale and black hairs. 5-8 mm. Europe, except Mediterranean, to Western Siberia **>** *Cheilosia vicina* Zetterstedt (= *Cheilosia nasutula* Becker)
   Note: for reference *C. albitarsis* may be used: it has a dusted posterior anepisternum.

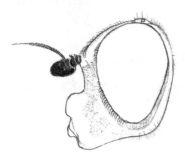

figure 129. *Cheilosia antiqua*, head of female (Verlinden).

6.a. Thoracic dorsum coarsely punctuate, interspaces 2-3 times the diameter of dots; head as figure 126, figure 127. Male: tergite 4 predominantly covered with black hairs; thoracic dorsum with long and short hairs. Female: thoracic dorsum with adpressed to semi-erect hairs, with black hairs at least in between wings; 6-8 mm. Palaearctic except North Africa 〉 *Cheilosia nigripes* Meigen

6.b. Thoracic dorsum finely punctuate, interspaces 5-10 times diameter of dots; head as figure 128, figure 129. Male: tergite 4 covered with pale or mixed pale and black hairs; thoracic dorsum with erect hairs of 1 length. Female: thoracic dorsum with entirely pale hairs forming a semi-erect short dense pile; 6-8 mm. Central Europe 〉 *Cheilosia antiqua* Meigen

7.a. Face shiny on facial knob and adjacent areas, frons shiny (as figure 128); thoracic dorsum mostly shiny, often with long black hairs mixed among the shorter pale ones 〉 8

7.b. Face and frons completely dusted, including the facial knob (figure 130, figure 131); thoracic dorsum almost completely dusted, with uniform dense, yellow pile of hairs only. 7-8 mm. Europe. 〉 *Cheilosa pubera* Zetterstedt
Jizz: dusted, brownish species.

8.a. Thoracic dorsum with pale hairs of a single length, sometimes with some longer black hairs at the back; sternites weakly dusted, (almost) shiny in centre; scutellar bristles sometimes lacking. Female head as figure 132. 6-9 mm. Central Europe, Northern Spain, Ireland. 〉 *Cheilosia ahenea* Von Roser
Jizz: dusted species, but face with more shiny parts than *C. pubera*.

8.b. Thoracic dorsum with short pale and long black hairs intermixed; sternites heavily dusted; scutellar bristles always present. Female head as figure 133. 7-8 mm. Northern and Central Europe. 〉 *Cheilosia sahlbergi* Becker

130.

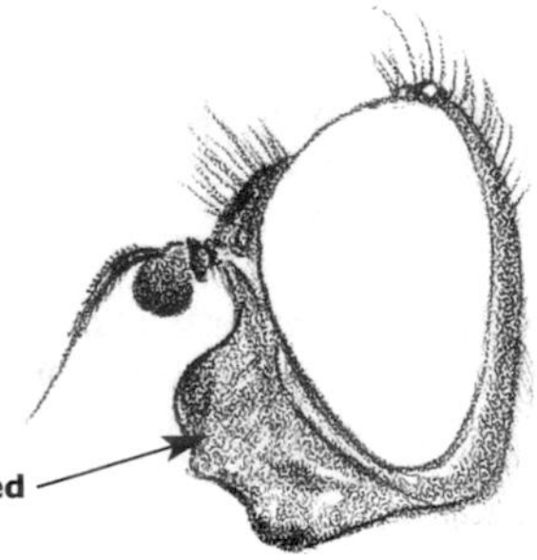

dusted

131.

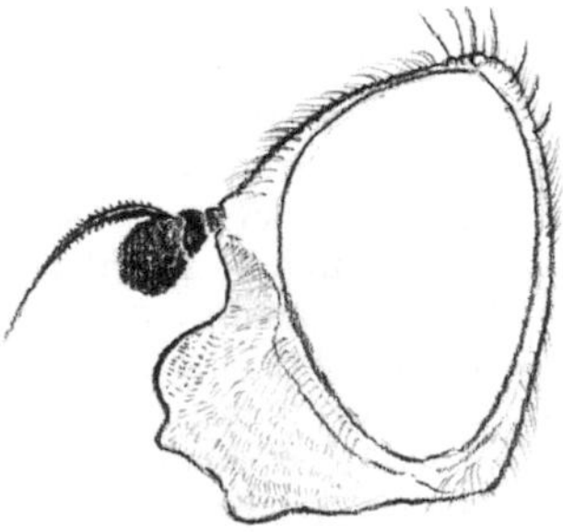

figure 130. *Cheilosia pubera*, head of male.
figure 131. *Cheilosia pubera*, head of female (Verlinden).

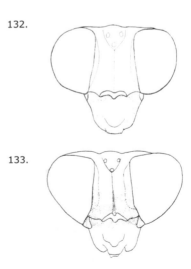

132.

133.

figure 132. *Cheilosia ahenea*, head of female.
figure 133. *Cheilosia sahlbergi*, head of female (Barkalov and Stahls, 1997).

Pagana group

1.a. Arista with hairs longer than the diameter of thickened basal part of arista (figure 134). Females: scutellum often, but not always, partly yellow ❯ 2

1.b. Arista bare or short-haired, hairs shorter than basal diameter of arista; scutellum black ❯ 6

2.a. Wing: base hyaline, brownish or blackish; face black. Female: scutellum black at least in front half ❯ 3

2.b. Wing: base yellow; face partly yellow. Female: scutellum yellow. 6-9 mm. Northern Europe, Siberia, Japan, North America ❯ *Cheilosia flavissima* Becker (= *Cheilosia pallipes* Loew)

3.a. Facial knob slender (figure 135): in the male nearly triangular, in the female only occupying 1/3 of the face viewed from above ❯ 4

3.b. Facial knob broad (figure 136): from above semi-circular running from eye margin to eye margin ❯ 5

4.a. Males and females. Male: thoracic dorsum: hairs black; head as figure 137. Female: humeri yellow, scutellum often with yellow hind part; frons without longitudinal scarf, head as figure 138. 6-9 mm. Northern and Central Europe to Mongolia. ❯ *Cheilosia longula* Zetterstedt
Jizz: dark, slender species. Tars 1 dark.

4.b. Females. Humeri and scutellum black; frons with longitudinal scarf (males with haired eyes). 5-8 mm. Europe to Siberia. ❯ *Cheilosia mutabilis* Fallén
Jizz: small species with elongated abdomen. Tars 1: middle segments often pale, others dark.

5.a. Antennae: 3rd segment bright red (figure 139, figure 140); anterior anepisternum with hairs along posterodorsal rim; thoracic dorsum with coarse punctation; 7-10 mm. Central and Southern Europe. ❯ *Cheilosia soror* Zetterstedt
Note: colour of 3rd antennal segment may vary!

135.

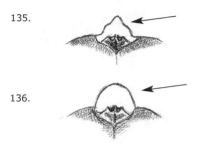

136.

figure 135. *Cheilosia longula*, top view of facial knob, pointed in front.
figure 136. *Cheilosia scutellata*, top view of facial knob, rounded in front (Verlinden).

137.

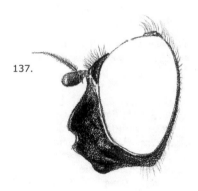

138.

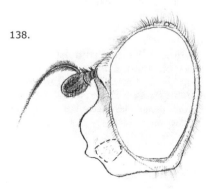

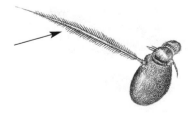

figure 134. *Cheilosia scutellata*, antenna (Verlinden).

figure 137. *Cheilosia longula*, head of male.
figure 138. *Cheilosia longula*, head of female (Verlinden).

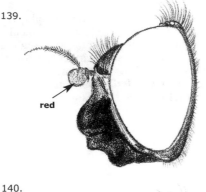

139.

red

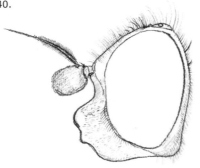

140.

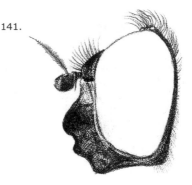

141.

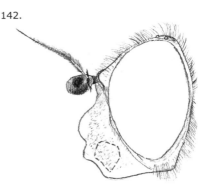

142.

figure 139. *Cheilosia soror*, head of male.
figure 140. *Cheilosia soror*, head of female (Verlinden).

figure 141. *Cheilosia scutellata*, head of male.
figure 142. *Cheilosia scutellata*, head of female (Verlinden).

5.b. Antennae: 3rd segment brown to black (figure 141, figure 142); anterior anepisternum without hairs along posterodorsal rim; thoracic dorsum with fine punctation. Male: thoracic dorsum predominantly with pale hairs. 7-10 mm. Palaearctic except North Africa. ❯ *Cheilosia scutellata* Fallén

6.a. Males and females. Femora predominantly black ❯ 7
6.b. Females. Femora completely pale; bristles on hind margin of scutellum absent. 7-11 mm. Central Europe ❯ *Cheilosia flavipes* Zetterstedt
Jizz: bare species with yellow legs. Males with haired eyes.

7.a. Wing clear; face does not protrude forwards (figure 146, figure 147) ❯ 8
7.b. Wing: cross veins in the middle of the wing darkened, often extending to a dark spot (figure 143); face protrudes forwards (figure 144, figure 145). 7-10 mm. Mountains of Central Europe, recently expanding into lowlands of

Northwest Europe where it appears in gardens. ❯ *Cheilosia caerulescens* Meigen

8.a. Hind margin of scutellum with bristles; eye margin below antennae with short white hairs; antennae: 3rd segment bright red ❯ 9
8.b. Hind margin of scutellum without bristles; eye margin below antennae with long white hairs; antennae: 3rd segment red, darkened above and sometimes at tip, sometimes turning more brownish; sternites dusted. 8-10 mm.

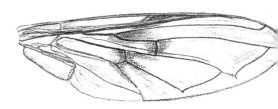

figure 143. *Cheilosia caerulescens*, wing (Verlinden).

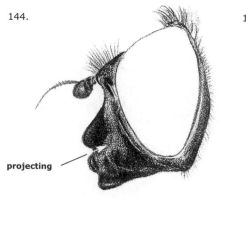

144.

projecting

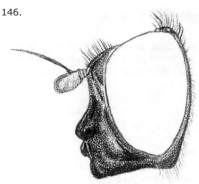

146.

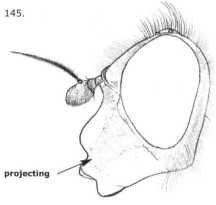

145.

projecting

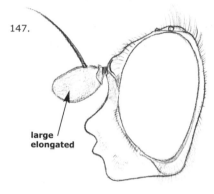

147.

large
elongated

figure 144. *Cheilosia caerulescens*, head of male.
figure 145. *Cheilosia caerulescens*, head of female (Verlinden).

figure 146. *Cheilosia pagana*, head of male.
figure 147. *Cheilosia pagana*, head of female (Verlinden).

Eastern Europe and southern parts of Central Europe, east to Afganistan, also in Sicily and Pyrenees **>** *Cheilosia laticornis* Rondani (= *Cheilosia latifacies* Loew)

Jizz: remniscent of *C. proxima* in having dusted sternites (but eyes bare).

9.a. Males and females, heads as figure 146, figure 147. Sternites shiny. **>** 10
9.b. Females. Sternites with thin silverish dust; 3rd antennal segment red, as long as wide without furrow on the outside; front and middle tarsae: 1st segments pale. 7-9 mm. Europe. **>** *Cheilosia uviformis* Becker

Jizz: antennae red but smaller than pagana. Males with haired eyes.

10.a Tars 1: central segments pale. Male: frons flat, not swollen in profile; 3rd antennal segment 1.5 times as long as wide. Female: 3rd antennal segment large and red, without furrow on the outside. Smaller: 5-8 mm. Holarctic **>** *Cheilosia pagana* Meigen

Jizz: females shiny with large, red antennae. Males more brownish.

10.b. Tars 1: central segments black to brown. Male: frons swollen in profile; 3rd antennal segment as long as wide. Female: 3rd antennal segment red, with a distinct furrow on the outside. Larger: 8-9 mm. Central Europe, Sweden. **>** *Cheilosia hercyniae* Loew

Variabilis group

1.a. Scutellar bristles long, at least as long as length of scutellum (figure 148); arista longer than twice the 3rd antennal segment > 2

1.b. Scutellar bristles less than 1/2 the scutellar length (figure 149); head as figure 150, figure 151, arista short, at most twice the length of 3rd antennal segment. 9-10 mm. Northern and Central Europe > *Cheilosia lasiopa* Kowarz (= *Cheilosia honesta* Rondani)

Jizz: specimens with dark legs end up here. Broad brownish species, arista twice as long as 3rd antennal segment.

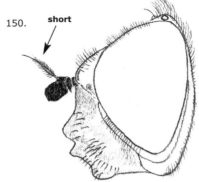

150.

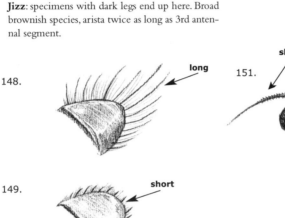

148.

149.

figure 148. *Cheilosia barbata*, scutellum with long bristles.
figure 149. *Cheilosia lasiopa*, scutellum with short bristles (Verlinden).

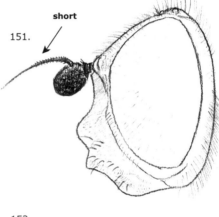

151.

152.

2.a. Males and females. Arista with hairs longer than diameter of basal thickened part; eyes with short hairs (figure 152). Male: halteres pale brown; larger: 9-12 mm. Europe to Western Siberia > *Cheilosia variabilis* Panzer

Jizz: large blackish species, black legs and antennae. Males patrol from large leaves along the forest edge.

2.b. Males. Arista with hairs shorter than basal diameter of arista; eyes with long hairs. Male: knob of halteres dark; smaller: 6-8 mm. Boreomontane: Northern Europe and mountains of Central Europe. > *Cheilosia melanopa* Zetterstedt

Jizz: small blackish species.

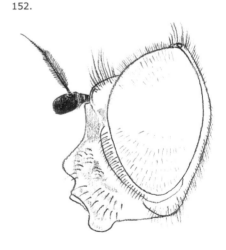

figure 150. *Cheilosia lasiopa*, head of male.
figure 151. *Cheilosia lasiopa*, head of female.
figure 152. *Cheilosia variabilis*, head of male (Verlinden).

Illustrata group

1.a. Wing without dark spot on middle; abdomen without dense, long hairs with the colour pattern described for *illustrata* ❯ 2

1.b. Wing with dark spot on middle (figure 153); abdomen with dense long hairs, white at front, black in middle and orange at tip; scutellar margin without bristles. 9-11 mm. Europe, east to Western Siberia ❯ *Cheilosia illustrata* Harris

Jizz: bumblebee mimic with orange abdominal tip. Resembles *Eristalis oestracea* and *Eriozona syrphoides*.

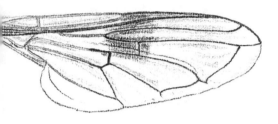

figure 153. *Cheilosia illustrata*, wing (Verlinden).

2.a. Scutellum with bristles at hind margin ❯ 3
2.b. Scutellum without bristles at hind margin ❯ 8

3.a. Scutellar bristles long, at least as long as length of scutellum (figure 148) ❯ 4
3.b. Scutellar bristles less than 1/2 the scutellar length (figure 149); arista short, at most twice the length of 3rd antennal segment. 9-10 mm. Northern and Central Europe ❯ *Cheilosia lasiopa* Kowarz (= *Cheilosia honesta* Rondani)
Jizz: broad brownish species.

4.a. Squamae whitish; thorax and abdomen: at least partly pale-haired. ❯ 5
4.b. Squamae blackish; thorax and abdomen black-haired; face and genae: in front view exceptionally broad (figure 174). Male: frons swollen. 8-9 mm. Northern Europe, east to Mongolia ❯ *Cheilosia morio* Zetterstedt
Jizz: very dark species with darkened wings.

5.a. Antennae: 3rd segment red, at most darkened at margins ❯ 6
5.b. Antennae: 3rd segment brown to black ❯ 7

6.a. Male: frons not swollen (figure 154). Females: antennae: 3rd segment large, as long as wide, arista with hairs longer than its diameter, arista longer than 3 times the length of the 3rd antennal segment (figure 155); thoracic dorsum with pale inclined hairs; margins of tergites uniformly haired 7-9 mm. Central Europe and mountains of Southern Europe ❯ *Cheilosia barbata* Loew ❯ 7

6.b. Male: frons strongly swollen (figure 156); abdomen: often with large dust spots, hairs on posteromedian part short and adpressed, long and erect laterally. Female: antennae: 3rd segment small, as long as wide; arista virtually bare, arista shorter than 3 times the length of the 3rd antennal segment (figure 157); thoracic dorsum with short silverish erect hairs mixed with black ones. 8-9 mm. Northern Europe and mountains of Central and Southern Europe, east to the Ural ❯ *Cheilosia frontalis* Loew

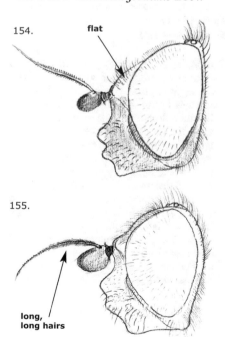

154.   flat

155.

long, long hairs

figure 154. *Cheilosia barbata*, head of male.
figure 155. *Cheilosia barbata*, head of female (Verlinden).

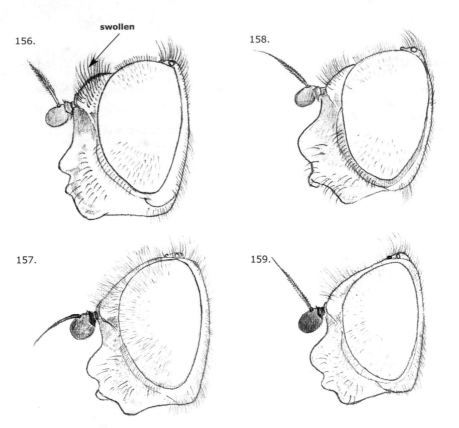

swollen

156.

157.

158.

159.

figure 156. *Cheilosia frontalis*, head of male.
figure 157. *Cheilosia frontalis*, head of female (Verlinden).

figure 158. *Cheilosia vulpina*, head of male.
figure 159. *Cheilosia vulpina*, head of female (Verlinden).

7.a. Males and females. Sternites dusted; arista less than 3 times the length of the 3rd segment; tibia black with pale bases; heads as figure 158, figure 159. Female: thoracic dorsum with grey to brown hairs of a single length, only a few black hairs; tergites with long, white hairs only on front corners of tergites; antennae: 3rd segment longer than wide. 8-10 mm. Midlands and mountains of Central and Southern Europe, east to the Ural **〉** *Cheilosia vulpina* Meigen (= *Cheilosia conops* Becker)

Jizz: somewhat bluish. Females: front corners of tergites appear whitish due to long hairs.

7.b. Females. Sternites shiny; thoracic dorsum with hairs of 2 lengths: shorter brown and many somewhat longer black hairs; margins of tergites uniformly haired with white, long hairs (legs of males completely black). 6-8 mm. Boreomontane: Northern Europe and mountains of Central Europe **〉** *Cheilosia melanopa* Zetterstedt

Jizz: blackish species with many dark hairs on thoracic dorsum.

8.a. Antenna: 3rd segment usually dark. Male: frons moderately shiny, with thin dust; head as high as broad in frontal view (figure 160); smaller: wing length 7.5 mm or less; tergite 3: grey spots nearly square or fused to form a band. Female: scutellar dorsum with short, often inclined hairs contrasting with longer hairs and bristles at scutellar hind margin; head in figure 161; smaller: wing length up to 6.5 mm; femur 3 usually black at base. 8-10 mm. Palaearctic **〉** *Cheilosia latifrons* Zetterstedt (= *Cheilosia intonsa* Loew)

Jizz: brownish species, eye margins very broad.

8.b. Antenna: 3rd segment usually reddish. Male: frons strongly grey dusted; head broader than wide in frontal view; larger: wing length 8-9 mm; tergite 3: grey spots triangular towards front corners. Female: scutellar dorsum with long, erect hairs, (almost) as long as hairs and bristles at scutellar margin; larger: wing length up to 7.5 mm; femur 3 usually orange at base. 7-8 mm. Central Europe, including Britain **〉** *Cheilosia griseiventris* Loew

Note: greyish species, see Stubbs and Falk (1983, 2002) for further details.

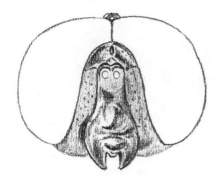

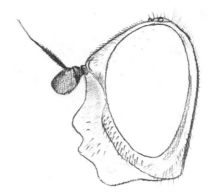

figure 160. *Cheilosia latifrons*, head of male (Verlinden).

figure 161. *Cheilosia latifrons*, head of female (Verlinden).

Impressa group

1.a. Males. Tars 1 and often 2 and 3: middle segments pale ❯ 2

1.b. Males and females. Tarsae: all segments black ❯ 3

2.a. Males: thoracic dorsum black-haired; tars 1: segment 5 square (figure 162); tergite 2 at the antero-lateral corner with black hairs (rarely absent); surstylus about 1.5 times as long as deep, with a wide carina and with a patch of microtrichia dorsally (figure 164); tergite 4: posterior margin with erect hairs, which may be mixed with some recumbent ones. Females with bare eyes. 7-9 mm. Palaearctic ❯ *Cheilosia albitarsis* Meigen

2.b. Males: thoracic dorsum with a narrow band of pale hairs anteriorly; tars 1: segment 5 trapezoid (figure 163); tergite 2 without black hairs antero-laterally; surstylus twice as long as deep, with a narrow carina and without a dorsal patch of microtrichia (figure 165); tergite 4: posterior margin with recumbent hairs only. Females with bare eyes. 7-9 mm. Central Europe ❯ *Cheilosia ranunculi* Doczkal

162.

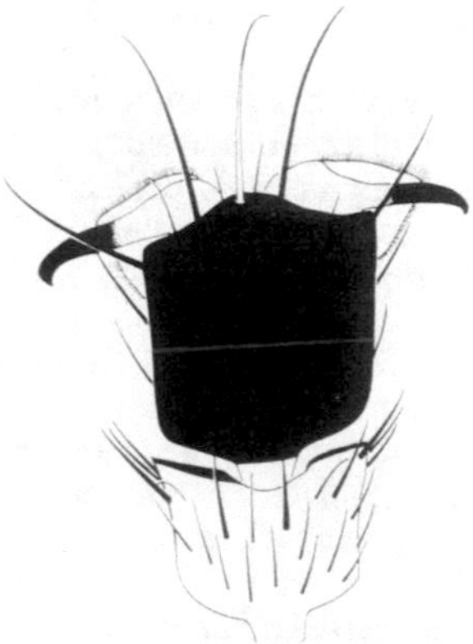

163.

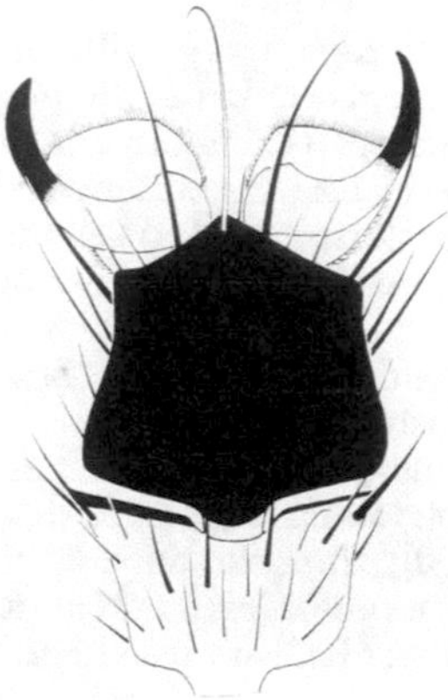

figure 162. *Cheilosia albitarsis*, tars 1 last segments.

figure 163. *Cheilosia ranunculi*, tars 1 last segments (Doczkal, 2000).

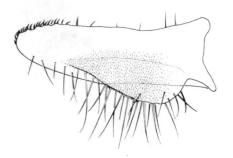

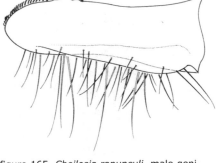

figure 164. *Cheilosia albitarsis*, male genitalia: surstylus, dorsal view (Doczkal, 2000).

figure 165. *Cheilosia ranunculi*, male genitalia: surstylus, dorsal view (Doczkal, 2000).

3.a. Wing: base pale: yellow to pale brown;
coxa 1 basally flattened; mouth edge not
projecting (figure 166, 167) **❯** 4

3.b. Wing: base hyaline to blackish; coxa 1
basally round; mouth edge projecting
(eg figure 168, 172, 176) **❯** 5

4.a. Coxa 1: with a hornlike projection at
the base of the outer side; facial knob
rounded, not strongly projecting (figure
166, figure 167); wing base yellow. 6-8
mm. Europe to East Russia **❯** *Cheilosia
impressa* Loew
Jizz: small, black species with yellow wing base
and abdomen much shorter than wings.

4.b. Coxa 1: without hornlike projection
at the base; facial knob rounded, even less
projecting than *C. impressa*; wing base
pale brown. 6-8mm. Northern Europe **❯**
Cheilosia naruska Haarto and Kerppola.

5.b. Mouth edge straight, although mouth
edge and facial knob do project forwards (figure 172, figure 173); tibia pale
at base, in rare cases completely black;
thoracic dorsum and scutellum shiny or
with broad longitudinal lines of dust **❯** 6

5.a. Mouth edge triangularly extended
downwards (figure 168, figure 169);
Legs including tibia completely black,
sometimes turning brownish at the base.
Male: abdomen with silverish spots (figure 170); thoracic dorsum uniformly
dusted, more or less in contrast to shiny
scutellum; eyes with pale hairs. 6-8 mm.
Northern and Central Europe **❯**
Cheilosia semifasciata Becker
Jizz: small size, with downwards projecting mouth
edge and (males) silver spots. Miniature *Ch. personata.*

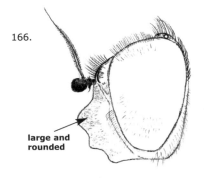

166.

large and
rounded

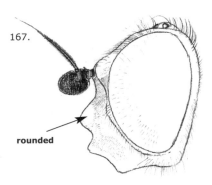

167.

rounded

figure 166. *Cheilosia impressa*, head of male.
figure 167. *Cheilosia impressa*, head of
female (Verlinden).

6.a. Tergites without grey spots, black
with dull patches in the middle **❯** 7

6.b. Tergites with grey spots (figure 171);
face below antennal implant gradually
sloping into facial knob (figure 172, figure 173). 7-8 mm. Early spring species.
Central Europe **❯** *Cheilosia fasciata*
Schiner & Egger
Jizz: greyish spots and tibia with pale bases, but
can be all black in the north. Larvae live in *Allium
ursinum*, the adults are often found near this plant.

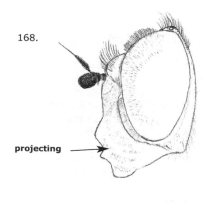

168.

projecting

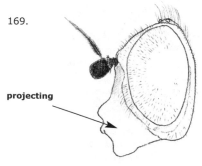

169.

projecting

figure 168. *Cheilosia semifasciata*, head of male.
figure 169. *Cheilosia semifasciata*, head of female (Verlinden).

7.a. Thoracic dorsum with 2 broad lines of dust; face and genae: in front view exceptionally broad (figure 174); mouth edge and facial knob not projecting forward. Female: abdomen oval. 7-10 mm. Northern Europe and the mountains of Central Europe **>** *Cheilosia morio* Zetterstedt
Jizz: blackish species, thorax and abdomen with black hairs. Wings hyaline.

7.b. Thoracic dorsum without lines of dust, shiny; face and genae: in front view not exceptionally broad; face below antennal implant angularly transfering into the projecting facial knob and mouth edge (figure 175, figure 176). Female: abdomen increases in width up to hind margin of tergite 3 then narrows abruptly (figure 177). 7-11 mm. Northern and Central Europe to Western Siberia **>** *Cheilosia carbonaria* Egger
Jizz: slightly bluish species; wing darkened; mouth edge and facial knob project squarely forwards.
Note: specimens of *C. naruska* without pale wing base may end up here, but mouth edge not projecting (see figure 166, 167).

170.

171.

figure 170. *Cheilosia semifasciata*, abdomen.
figure 171. *Cheilosia fasciata*, abdomen (Verlinden).

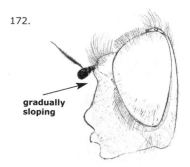

172.

gradually sloping

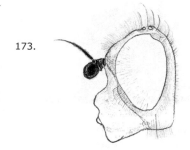

173.

figure 172. *Cheilosia fasciata*, head of male.
figure 173. *Cheilosia fasciata*, head of female (Verlinden).

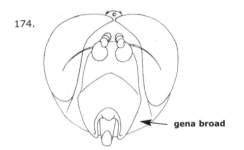

174.

gena broad

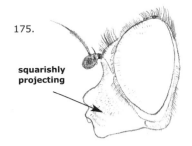

175.

squarishly
projecting

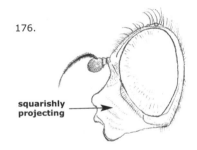

176.

squarishly
projecting

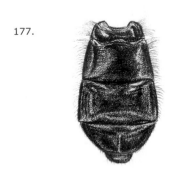

177.

figure 174. *Cheilosia morio*, head, front view
(after Van der Goot, 1981).
figure 175. *Cheilosia carbonaria*, head of
male.
figure 176. *Cheilosia carbonaria*, head of
female.
figure 177. *Cheilosia carbonaria*, abdomen of
female (Verlinden).

Proxima group

1.a. Males ❭ 2
1.b. Females ❭ 8

2.a. Frons completely dusted; lunula yellow; antennae: 3rd segment at least
basoventrally clear reddish, often completely reddish ❭ 3
2.b. Frons shiny, sometimes dusted at eye
margin; lunula brown to black; antennae
in mature specimens brown to black, at
most turning paler basoventrally (only
in *Ch. velutina* partly red) ❭ 4

3.a. Face mostly dusted (figure 178); ster-
nites with dull brownish dust. 9-10 mm.
Central Europe ❭ *Cheilosia rufimana*
Becker
Jizz: broad species with bronze-green sheen.
3.b. Face mostly shiny (figure 179); stern-
ites with semi-shiny silverish dust. 7-9
mm. Europe. ❭ *Cheilosia uviformis*
Becker (= *Cheilosia argentifrons* Hellen)
Jizz: slender, blackish species.

4.a. Thoracic and scutellar dorsum with
hairs of the same length, only some hairs
on the thoracic dorsum before the
scutellum longer ❭ 5
4.b. Thoracic and especially scutellar dorsum with hairs of 2 lengths, longer black
hairs and shorter pale to dark hairs ❭ 6

5.a. Smaller species: 6-8 mm; antennal
segment 3 often pale with top margin
darkened; facial knob and mouth edge
poorly developed (figure 180); bristles at
scutellar margin as long as the scutellar
length; tarsae pale for some distance.
Palaearctic. ❭ *Cheilosia velutina* Loew
Jizz: at first similar to *C. proxima*, but facial knob
virtually lacking.
5.b. Larger species: 9-11 mm; antennal
segment 3 often black with base turning
pale; facial knob and mouth edge well-
developed; bristles at scutellar margin
longer than scutellar length; tarsae black
above. Northern and Central Europe to
Eastern Russia ❭ *Cheilosia gigantea*
Zetterstedt
Jizz: large species, with pale parts on legs, at first
sight not unlike *C. variabilis* (has black legs).

178.

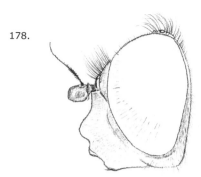

figure 181. *Cheilosia ingerae*, head of male
(Nielsen and Claussen, 2001).

179.

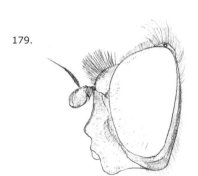

7.a Anterior anepisternum bare along posterodorsal rim (figure 183); thoracic dorsum without field of black bristles before
scutellum; thoracic dorsum finely punctuate; head as figure 182. 7-9 mm.
Northern and Central Europe to Eastern
Russia > *Cheilosia proxima* Zetterstedt

7.b. Anterior anepisternum with hairs along
posterodorsal rim (figure 184); thoracic
dorsum with a field of small black hairs
just before the scutellum; thoracic dorsum
coarsely punctuate. 7-9 mm. Central
Europe > *Cheilosia aerea* Dufour

180.

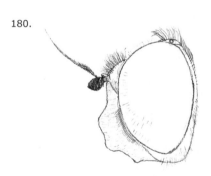

figure 178. *Cheilosia rufimana*, head of male.
figure 179. *Cheilosia uviformis*, head of male.
figure 180. *Cheilosia velutina*, head of male
(Verlinden).

8.a. Anterior anepisternum bare along
posterodorsal rim (figure 183); thoracic
dorsum finely punctuate, with hairs that
may be erect and partly black > 9

8.b. Anterior anepisternum with hairs
along posterodorsal rim (figure 184);
thoracic dorsum coarsely punctuate;
thoracic dorsum: hairs all pale, semi-
erect; facial knob well-developed, but
mouth edge hardly elevated. 7-9 mm. >
Cheilosia aerea Dufour

9.a. Punctuation of lower part of frons
and thorac dorsum of equal density,
interspaces shiny (then body length 8
mm or more) or finely chagrinated;
bristle on hind rim of posterior anepisternum may be present; facial knob and
mouth edge well-developed > 10

9.b. Punctuation on frons far less dense
than on thoracic dorsum, with shiny
interspaces; bristle on hind margin of
posterior anepisternum always absent;
facial knob and mouth edge hardly
developed (figure 185); tergite 3: patch of
dust divided in 2, not reaching hind margin; 6-8 mm > *Cheilosia velutina* Loew

6.a. Margin of upper calypter with short
pale bristles, frons not swollen > 7

6.b. Margin of upper calypter with short
black or dark brown bristles; frons
slightly swollen (figure 181). 7-9.5 mm.
Subarctic parts of Northern Europe. >
Cheilosia ingerae Nielsen & Claussen

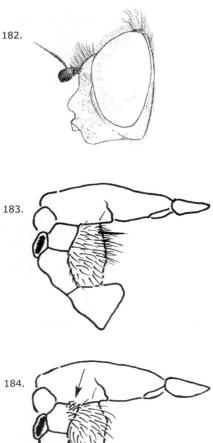

182.

183.

184.

figure 182. *Cheilosia proxima*, head of male (Verlinden).
figure 183. *Cheilosia proxima*, thorax, side view.
figure 184. *Cheilosia aerea*, thorax, side view (Reemer in Smit, Reemer and Renema, 2001).

10.a. Femur 3 without a row of black bristles at the tip on the lower side (at most a few at the tip: *Ch. gigantea*); lunula brown to black; antenna: 3rd segment brown to black, often with base turning orange; frons not dusted, punctuate with interspaces shiny or chagrinated; tergites with black or blue sheen ❯ 11

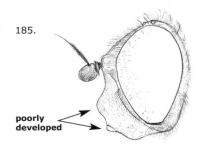

185.

poorly developed

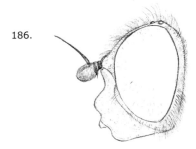

186.

figure 185. *Cheilosia velutina*, head of female.
figure 186. *Cheilosia rufimana*, head of female (Verlinden).

10.b Femur 3 with a row of stout, black bristles at the tip on the lower side; lunula yellowish; antenna: 3rd segment bright orange, darkened at top margin (figure 186); thoracic dorsum with pale hairs only, these hairs erect or semi-erect; frons weakly dusted, punctuate with interspaces finely chagrinated; tergites with copper sheen. 9-10 mm. ❯ *Cheilosia rufimana* Becker

11.a Posterior anepisternum: more than anterior 1/3 grey dusted; thoracic dorsum with black hairs intermixed with pale ones, these black hairs may extend above the others, and black hairs may dominate; frons relatively narrow, length/width ratio 1.4-1.7 ❯ 12

11.b Posterior anepisternum: at most anterior 1/3 thinly dusted; thoracic dorsum with short, erect pale hairs, sometimes with single longer black ones intermixed; femur 3: basal 2/3 with anteroventral hair fringe as long as diameter of femur; frons broad, length/width ratio 1.2-1.4 (figure 187). 7-9.5 mm. ❯ *Cheilosia ingerae* Nielsen & Claussen

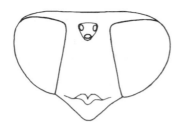

figure 187. *Cheilosia ingerae*, head of female (Nielsen and Claussen, 2001).

2.a. Eyes: hairs all white to pale brown ❯ 3
2.b. Eyes: at least the hairs on upper 1/2 dark brown to black ❯ 11
Note: take care, strong white light reflects off black hairs, causing them to appear white!

3.a. Femur 3: underside densely covered with many black bristles ❯ 4
3.b. Femur 3: underside not covered with bristles, at most with a row of strong hairs ❯ 7

4.a. Males and females. Eyes: hairs short, equal to hairs on eye rims; bristles at hind margin scutellum and postalar calli black or black and yellow equally intermixed ❯ 5
4.b. Females. Eyes: hairs long, longer than hairs at eye rims; bristles at hind margin scutellum and postalar calli all yellow (seldom a few black bristles); tibia 3 pale with black ring; thorax and abdomen with pale hairs; head as figure 189. 9-12 mm. Central Europe ❯ *Cheilosia lenis* Becker
Jizz: large with reddish antennae.

12.a. Small species: 7-9 mm; frons: interspaces between dots finely chagrinated; femur 3: basal 2/3 with the anteroventral hair fringe long, often longer than diameter of femur; tergite 3: dust patch not divided, reaches hind margin; bristle on hind rim of posterior anepisternum present; head as figure 188. ❯ *Cheilosia proxima* Zetterstedt
12.b. Large species: 9-11 mm; frons: interspaces shiny; femur 3: without anteroventral hair fringe, at most a few single long hairs; tergite 3: dust patch may be divided by a shiny line, not reaching hind margin; bristle on hind rim of posterior anepisternum may be absent ❯ *Cheilosia gigantea* Zetterstedt

Bergenstammi group

1.a. Tibia pale at both ends ❯ 2
1.b. Tibia only pale at their bases, black at tip ❯ *C.impressa*-group (p. 65)

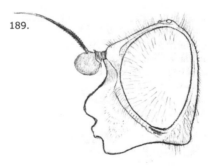

189.

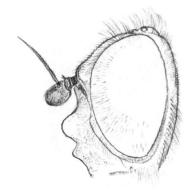

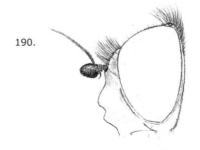

190.

figure 188. *Cheilosia proxima*, head of female (Verlinden).

figure 189. *Cheilosia lenis*, head of female.
figure 190. *Cheilosia mutabilis*, head of male (Verlinden).

5.a. Males and females. Antennae: 3rd segment as long as wide, often entirely bright orange and never entirely black; facial knob strong and projecting forward; larger species: (7-)9-12 ❭ 6

5.b. Males. Antennae: 3rd segment 1.5 times as long as wide, often entirely black and never entirely bright orange; facial knob moderate (figure 190); tars 1: metatars and last segments dark with contrasting pale middle segments (figure 191); smaller species: 6-8 mm. Northern and Central Europe to Western Siberia ❭ *Cheilosia mutabilis* Fallén

6.a. Tibia 3 pale with black ring; tarsae: most segments pale; abdomen elongate (males) or oval (females); head with facial knob and mouth edge moderately projecting figure 192, figure 193. 9-12 mm. Northern and Central Europe ❭ *Cheilosia bergenstammi* Becker
Jizz: large and brownish, head dark with reddish antennae.

6.b. Tibia and tars 3 dark except for bases of tibia; head with strongly projecting facial knob and mouth edge (figure 175, figure 176). Females: abdomen increases in width up to tergite 3 and then strongly narrows (figure 177). (7-) 9-11 mm. Northern and Central Europe to Western Siberia ❭ *Cheilosia carbonaria* Egger
Jizz: black species with dark legs, facial knob large. Males similar to *C. cynocephala*, but thoracic dorsum without bluish sheen, legs darker and facial knob larger.

7.a. Males and females. Arista: bare or hairs shorter than basal diameter of arista ❭ 8

7.b. Males. Arista with hairs equal to or longer than the basal diameter of arista; antennae: 3rd segment 1.5 times as long as

wide, often black, sometimes orange basally, but never bright orange (figure 194); tars 1: commonly metatars and last segment dark with contrasting pale middle segments (figure 191). 6-8 mm. Northern and Central Europe to Western Siberia ❭ *Cheilosia mutabilis* Fallén

8.a. Antennae: 3rd segment elongate, about 1.5 times as long as wide; antennae: colour very variable, black to red ❭ 9

8.b. Antennae: 3rd segment round, as long as wide; antennae always red ❭ 10

9. a. Claws on tarsae pale orange on basal, black on apical 1/2; head as figure 195, figure 196. Male: length of contact line of eyes about 1.5 times the length of the frons; frons not swollen; angle of eyes about 90°; hypandrium: dorsal prominence of the gonostylus long and erect. Female: dust spots on frons present, connected to dusting on face; apex of femora pale over considerable distance; tergite 5 mostly undusted. 6-8 mm. Northern and Central Europe ❭ *Cheilosia urbana* Meigen (= *Cheilosia praecox* Zetterstedt)

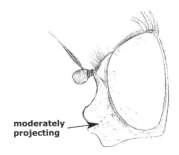

moderately projecting

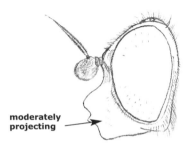

moderately projecting

figure 192. *Cheilosia bergenstammi*, head of male.

figure 193. *Cheilosia bergenstammi*, head of female (Verlinden).

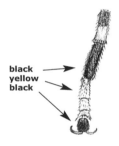

black
yellow
black

figure 191. *Cheilosia mutabilis*, tars 1 (Verlinden).

9.b. Claws on tarsae dark, brown at base. Male: length of contact line of eyes 0.9-1.2 times the length of the frons; frons swollen; angle of eyes 90-100°; hypandrium: dorsal prominence of the gonostylus short and reclinate. Female: dust spots of frons, if present, not connected to dusting on face; apex of femora pale for a short distance; tergite 5 mostly weakly dusted. 6-8 mm. Central Europe **> Cheilosia psilophthalma** Becker

194.

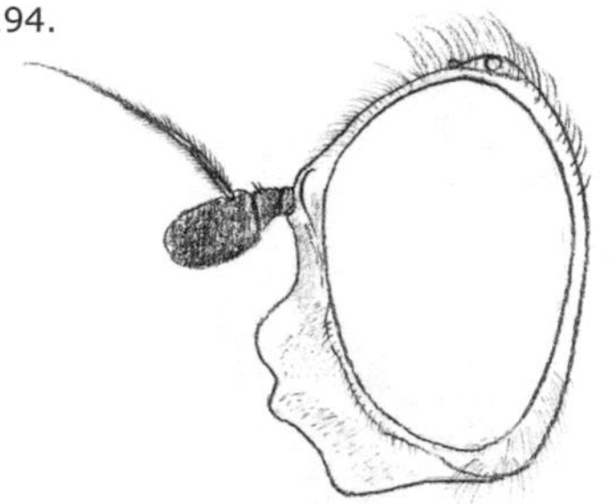

195.

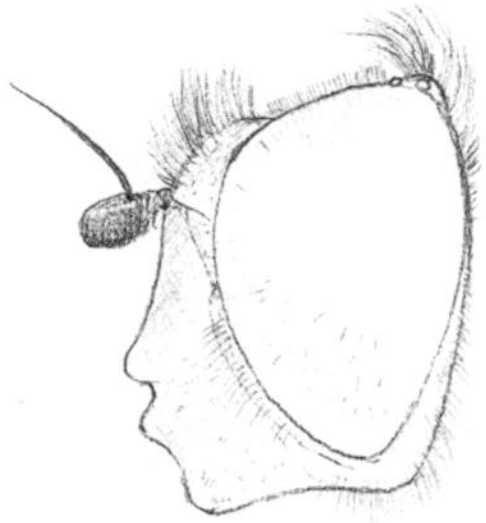

196.

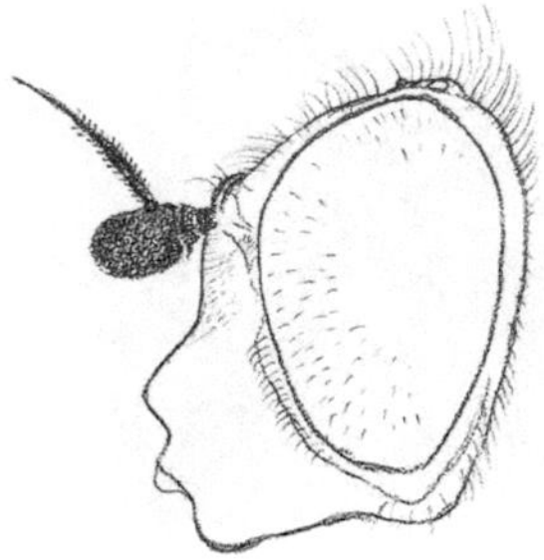

figure 194. *Cheilosia mutabilis*, head of female.
figure 195. *Cheilosia urbana*, head of male.
figure 196. *Cheilosia urbana*, head of female (Verlinden).

10.a. Sternites shiny. Male: frons shiny. Female: eyes densely haired; abdomen round or elongated. Larger species: 8-12 mm **>** *fraterna*-subgroup (p. 79)
10.b. Sternites weakly dusted. Male: frons dusted. Female: eyes almost bare; abdomen much longer than wide. Smaller species: 7-9 mm **> Cheilosia uviformis** Becker
Note: should key out in the proxima group, but some have very weak sternite dusting.

11.a. Males and females. Katepisternum: upper and lower hair patches separated; eyes: hairs on middle of eye equal to hairs on eye rims (seldom eye hairs somewhat longer than hairs on eye rim), either all black or white on lower 1/4 of eye; a number of smaller species: 5-9 mm, but one larger species: 11-13 mm; **>** 12
11.b. Males. Katepisternum: upper and lower hair patches joined along front; eyes: hairs on middle of eye long, exceeding length of hairs on eye rims and all black; 1st 2 segments of tars 1 and 2 pale; head as figure 197; larger species: 9-12 mm. Central Europe **> Cheilosia lenis** Becker
Jizz: large with reddish antennae.

12.a. Wing hyaline; thoracic dorsum blackish or bronze. Male: tergite 5 shiny or weakly dusted, almost as glossy as tergite 4; tergites 3 and 4 with hairs at side margin yellowish or only partially black; hairs on thoracic dorsum predominantly yellow, in some species with black hairs between wings **>** 13
Note: the species following 12a are hard to separate due to large intraspecific variability.
12.b. Wing darkened in apical half (distinct in females but indistinct in males); thoracic dorsum with bluish sheen; heads as figure 198, figure 199. Male: tergite 5 clearly dusted in contrast with shiny segment 4; tergites 3 and 4 with hairs at side margin entirely or predominantly black; hairs on thoracic dorsum predominantly black. 7-8 mm. Northern and Central Europe to Western Siberia **> Cheilosia cynocephala** Loew
Jizz: black with bluish sheen, darkened wings and dark hairs. Like *C. carbonaria*, but legs paler and smaller facial knob.

13.a. Thoracic dorsum and abdomen with pale and black hairs, the extent of black hairs can be reduced to a few between the wings ❯ 14
13.b. Thoracic dorsum and abdomen with only pale hairs ❯ 16

14.a. Eyes densely haired all over, with white hairs on lower part (figure 200); thoracic dorsum without distinct band of black hairs between wings; facial knob less protruding, not extending to or beyond tip of antennae; legs less dark, base and apex of tibiae pale for 1/4 or more; smaller: 5-7 (8) mm ❯ 15

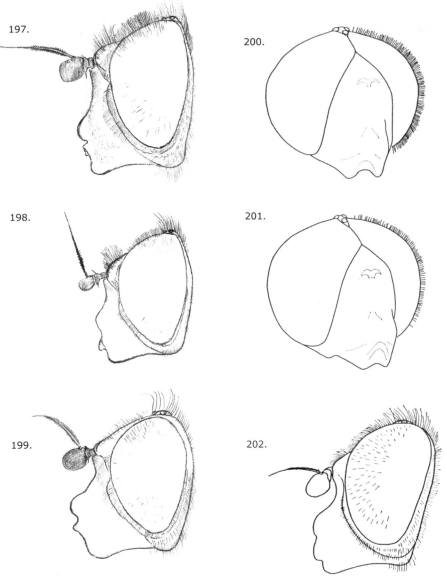

197.

198.

199.

200.

201.

202.

figure 197. *Cheilosia lenis*, head of male.
figure 198. *Cheilosia cynocephala*, head of male.
figure 199. *Cheilosia cynocephala*, head of female (Verlinden).

figure 200. *Cheilosia vernalis*, head of male.
figure 201. *Cheilosia sootryeni*, head of male.
figure 202. *Cheilosia sootryeni*, head of male (Nielsen, 1970).

14.b. Eyes densely haired on upper part, but on lower part hairs very sparse to almost absent, brown (figure 201); thoracic dorsum at least with some black hairs between wings, which may extend to a black band; facial knob strong, extends to or beyond antennae in side view (figure 202); legs relatively dark: about basal 1/3 and apical 1/6 of tibiae pale; 7-9 mm. Norway **> *Cheilosia sootryeni*** Nielsen

15.a. Scutellum: bristles on hind margin shorter than scutellum length (figure 203); tibia 3 with dark ring longer than pale parts (figure 204); head as figure 205, figure 206. 5-7 mm. Europe to Eastern Russia **> *Cheilosia vernalis*** Fallén

15.b. Scutellum: bristles on hind margin as long as scutellum length (figure 207); tibia 3 with dark ring shorter than pale parts (figure 208); heads as figure 209, figure 210. 5-7 mm. Central Europe **> *Cheilosia rotundiventris*** Becker

Note: conspecific to *C. vernalis* according to some authors.

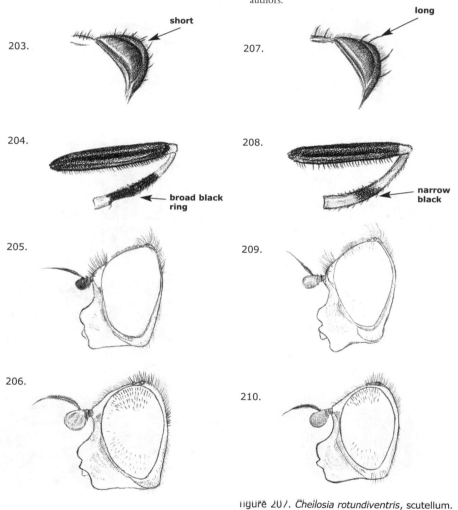

203.

short

204.

broad black ring

205.

206.

207.

long

208.

narrow black

209.

210.

figure 203. *Cheilosia vernalis*, scutellum.
figure 204. *Cheilosia vernalis*, leg 3.
figure 205. *Cheilosia vernalis*, head of male.
figure 206. *Cheilosia vernalis*, head of female (Verlinden).

figure 207. *Cheilosia rotundiventris*, scutellum.
figure 208. *Cheilosia rotundiventris*, leg 3.
figure 209. *Cheilosia rotundiventris*, head of male.
figure 210. *Cheilosia rotundiventris*, head of female (Verlinden).

16.a. Males and females (figure 211, figure 212). Smaller species: 6-7 mm; antennae: 3rd segment brownish; tars 2: basal segments pale above. Central Europe ❯ *Cheilosia ruficollis* Becker

16.b. Males (figure 213). Larger species: 11-13 mm; antennae: 3rd segment reddish; tars 2: basal segments dark above. 11-13 mm. Central Europe and elevated parts of Southern Europe ❯ *Cheilosia bracusi* Vujic & Claussen

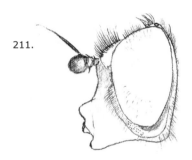

211.

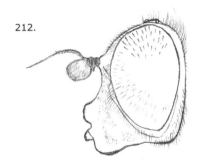

212.

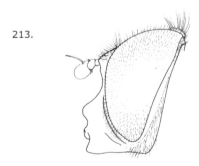

213.

figure 211. *Cheilosia ruficollis*, head of male (Verlinden).
figure 212. *Cheilosia ruficollis*, head of female (Verlinden).
figure 213. *Cheilosia bracusi*, head of male (Vujic and Claussen, 1994b).

Canicularis group

1.a. Antennae: 3rd segment brown to black ❯ 2
1.b. Antennae: 3rd segment reddish ❯ 3

2.a. Tarsae: all segments black; head as figure 214. 11-12 mm. Early spring species. Northern and Central Europe, higher parts of Southern Europe, Western Siberia, Oriental region ❯ *Cheilosia grossa* Fallén
Jizz: bee mimic with long but not dense hairs on body.

2.b. Tars 2 and often tars 1 pale over a considerable length. 9-11 mm. Northern Europe ❯ *Cheilosia alpina* Zetterstedt
Jizz: bee mimic with long but not dense hairs on body.

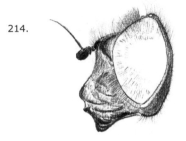

214.

215.

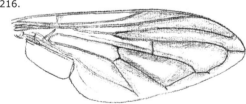

216.

figure 214. *Cheilosia grossa*, head of male.
figure 215. *Cheilosia chrysocoma*, wing.
figure 216. *Cheilosia nebulosa*, wing (Verlinden).

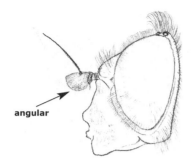

angular

figure 217. *Cheilosia chrysocoma*, head of male (Verlinden).

3.a. Wing with distinct dark patterns, especially on cross veins in the middle (figure 215, figure 216) **>** 4
3.b. Wing hyaline or with diffuse yellow-ish or brownish tint **>** 6

4.a. Abdomen with long yellowish hairs; antennae: 3rd segment rounded at tip **>** 5
4.b. Abdomen with long, dense reddish hairs; antennae: 3rd segment angular at tip (figure 217); wing: cross veins dark-ened (figure 215). 9-11 mm. Northern and Central Europe, east to Siberia **>** *Cheilosia chrysocoma* Meigen
Jizz: like an *Osmia* (Apidae) at first sight, with its dense reddish hairs and erratic flight.

5.a. Wing with central cross veins dark-ened and typically with dusky clouds on anterior part (figure 216); antennae: 3rd segment pale with dark tip (figure 218, figure 219). Smaller: 9-10 mm. Northern and Central Europe **>** *Cheilosia nebulosa* Verrall (= *Cheilosia langhofferi* Becker)
Jizz: quite like a small *C. albipila*: antennae reddish, abdomen with long hairs, tibia 3 completely pale.
5.b. Wing with a single dark spot at mid-dle; antennae: 3rd segment completely pale. Larger: 10-14 mm. Mountains of Central and Southern Europe to Western Siberia **>** *Cheilosia subpic-tipennis* Claussen
Jizz: large, shiny species with long pale hairs.

6.a. Anepisternum: hairs wavy at their top 1/4 (figure 220), many long hairs on the femora and sternites also wavy at their top **>** 7

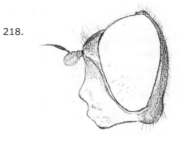

218.

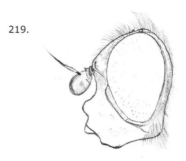

219.

figure 218. *Cheilosia nebulosa*, head of male.
figure 219. *Cheilosia nebulosa*, head of female (Verlinden).

6.b. Anepisternum: hairs straight at their top, at most a few wavy at their very end (figure 221), long hairs on femora and sternites with straight tips **>** 9

7.a. Larger species: 12-15 mm; tarsae largely pale, metatars and last segment may be darkened; femora black; hairs on thorax and abdomen shorter and vary-ing in colour between white and red-dish; tergites with copper sheen **>** 8

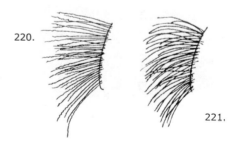

220.

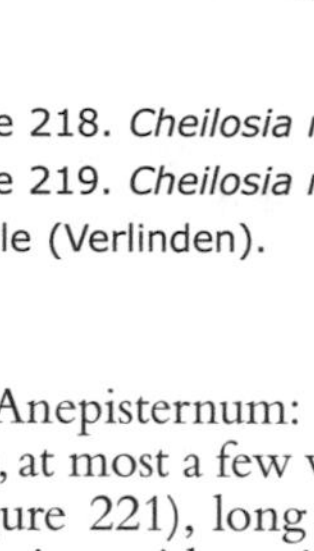

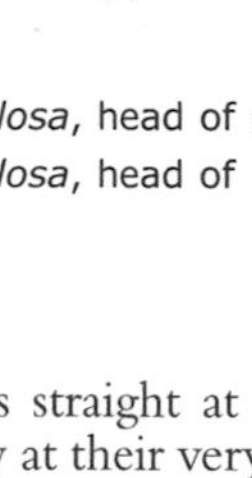

221.

figure 220. *Cheilosia himantopus*, hairs on anepisternum.
figure 221. *Cheilosia orthotricha,* hairs on anepisternum (Vujic and Claussen, 1994a).

7.b. Smaller species: 8-12 mm; heads as figure 222, figure 223. Male: tarsae dark; femora mostly dark. Female: tarsae yellow, last segments often darkened, femora completely yellow. Both: hairs on thorax and abdomen long and reddish; tergites with purple sheen. Northern and Central Europe to Western Siberia ❯ *Cheilosia albipila* Meigen

Jizz: body more (male) or less (female) bicoloured: thorax greenish, abdomen purple, covered with long reddish hairs, tibia completely pale.

Note: if arista yellow (-brown) instead of black and frons strongly swollen in the male, check *Ch. nebulosa*, which may lack the brown pattern in the wing.

8.a. Arista: hairs about 1/2 the basal width of the arista (figure 224). Male: last segment of tarsae often dark. Female: tergite 3: hind margin with interrupted band of adpressed hairs; tergite 4: hind 1/3 with sparse, erect hairs. 11-13 mm. Central Europe ❯ *Cheilosia himantopa* Panzer

Jizz: large species, last segments of tarsae darkened.

8.b. Arista: hairs longer than the basal width of the arista (figure 225); head as figure 226, figure 227. Male: last segment of tarsae often pale. Female: tergite 3: hind margin with a band of semi-erect hairs; tergite 4: hind third with dense, semi-erect hairs. 11-13 mm. Central Europe to Japan ❯ *Cheilosia canicularis* Panzer

Jizz: large species, last segments of tarsae often pale.

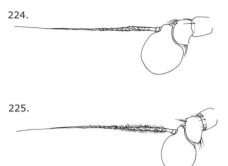

figure 224. *Cheilosia himantopus*, antenna.
figure 225. *Cheilosia canicularis*, antenna
(Stuke and Claussen, 2000).

9.a. Tars 1 darkened above, at most the metatars turning pale; arista 3-4 times the length of 3rd antennal segment; smaller species: 7-12 mm ❯ 10

9.b. Tars 1 completely pale, exceptionally with metatars and top segment darkened; arista long, more than 4 times the length of 3rd antennal segment; heads as figure 228, figure 229; larger species: 12-15 mm. Central Europe ❯ *Cheilosia orthotricha* Vujic & Claussen

Jizz: large species, tibia 3 almost completely pale.

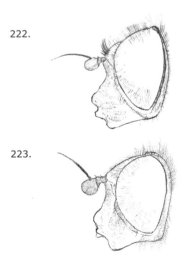

figure 222. *Cheilosia albipila*, head of male.
figure 223. *Cheilosia albipila*, head of female
(Verlinden).

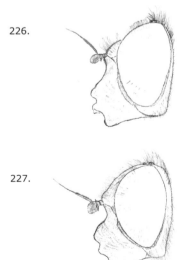

figure 226. *Cheilosia canicularis*, head of male.
figure 227. *Cheilosia canicularis*, head of
female (Verlinden).

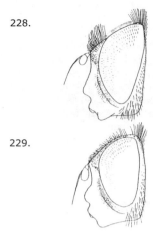

228.

229.

figure 228. *Cheilosia orthotricha*, head of male.

figure 229. *Cheilosia orthotricha*, head of female (Vujic and Claussen, 1994a).

10.a. Tibia 3 with a dark ring or a dark patch on the anterior side. Male: femur 3 with dense black bristles or strong black hairs ventrally; abdomen with long, erect hairs. Female: abdomen round, as long as wide **>** *fraterna*-subgroup

10.b. Tibia 3 entirely yellow. Male: femur 3 without black bristles ventrally **>** 11

11.a. Males and females. Tars 1 and 2 mainly black above; frons: hairs all black. 7-9mm. Southern Finland. **>** *Cheilosia alba* Vujic and Claussen

11.b. Males only. Tars 1 and 2: at least three basal segments yellow above; frons: pale hairs present among the black; head as figure 230. 7-11 mm. Central Europe **>** *Cheilosia flavipes* Zetterstedt

Note: female has bare eyes.

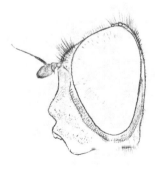

figure 230. *Cheilosia flavipes*, head of male (Verlinden).

Fraterna-subgroup

The subgroup is recognized because its species may or may not have bristles at the scutellar hind margin and postalar calli and may key out to both the *bergenstammi* and *canicularis* subgroups.

1.a. Tibiae mainly pale, tibia 3 pale or pale with a black band over less than 1/2 its length. Male: pregenital segment pale-haired; thoracic dorsum completely pale-haired or with shorter pale and longer black hairs; Female: thoracic and scutellar dorsum with pale hairs only **>** 2

1.b. Tibiae mainly black, tibia 3 with a black band over more than 1/2 its length; Thoracic dorsum with many black hairs mixed with pale hairs, which may dominate before the suture. Male: pregenital segment black-haired; eyes: black hairs. 9-11 mm. Poland, the Alps **>** *Cheilosia melanura* Becker

2.a. Arista uniformly dark, brown to black at base (figure 231, figure 232); bristles on scutellar hind margin present or absent, bristles on postalar calli present; eyes: hairs pale to black **>** 3

2.b. Arista reddish at base, dark on remaining parts (figure 233, figure 234); tibia 3 pale with black ring below middle, which may be reduced; bristles on scutellar hind margin and postalar calli absent; eyes: hairs pale to brown. 8-11 mm. Northern and Central Europe to Western Siberia **>** *Cheilosia chloris* Meigen

3.a. Large species: 8-13mm; eyes: hairy all over or bare on lower part; tibia 3: pale with black patch or pale with black ring. Female: hairs on thoracic dorsum short -**>** 4

3.b. Small species: 7-8mm. Male: eyes hairy all over and tibia 3 with black ring (as the much larger *Cheilosia bracusi* of central Europe). Females: thoracic dorsum with extremely short hairs; eyes bare on lower part; tibia 3 with black ring. Finland only **>** *Cheilosia reniformis* Hellén

4.a. Eyes: virtually bare on lower part; tibia 3 pale with dark patch halfway, which may be reduced; bristles on scutellar hind margin present for 80% of specimens, on postalar calli present for 99% of all specimens; femur 3: lower surface at the tip without bristles, but with strong black hairs; head as figure 235, figure 236. Male: eyes: hairs all pale. Female: hairs on thoracic dorsum short and erect; abdomen round, as long as broad; eyes: hairs all pale. 8-10 mm. Finland south to the Pyrenees, east into Siberia › *Cheilosia fraterna* Meigen

4.b. Eyes: hairy all over; tibia 3 pale with dark ring below middle; femur 3: lower surface has black bristles at the tip; head as figure 213. Male: bristles on scutellar hind margin absent (but strong hairs may look like bristles), bristles on postalar calli present; eyes: hairs black on upper parts, brown on lower parts. Female: bristles on scutellar margin and postalar calli absent; abdomen longer than broad; hairs on thoracic dorsum short and adpressed; eyes: hairs all pale. 11-13 mm. Eastern Central Europe and elevated parts of Southern Europe › *Cheilosia bracusi* Vujic and Claussen

CHRYSOGASTER

Introduction

Chrysogaster are medium-sized Syrphid flies. They are black in colour, with a green or purple sheen. The flies can be found on flowers, and prefer umbellifers and buttercups. The larvae are aquatic and semi-aquatic. Larvae of *Chrysogaster solstitialis* are reported to live in damp depressions in the forest soil where water seeps through and in deposits of organic material near forest streams. They live in the upper 1-2 cm of the soil amongst organic material.

The key is based on Van der Goot (1981), Maibach and Goeldlin de Tiefeneau (1994) and Maibach and Goeldlin de Tiefeneau (1995).

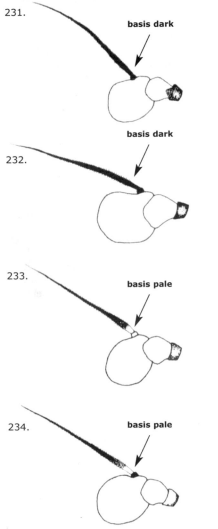

231.

basis dark

basis dark

232.

233.

basis pale

234.

basis pale

figure 231. *Cheilosia bracusi,* antenna male.
figure 232. *Cheilosia bracusi,* antenna female.
figure 233. *Cheilosia chloris,* antenna male.
figure 234. *Cheilosia chloris,* antenna female (Vujic and Claussen, 1994b).

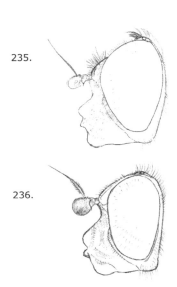

235.

236.

figure 235. *Cheilosia fraterna,* head of male.
figure 236. *Cheilosia fraterna,* head of female (Verlinden).

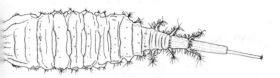

figure 237. *Chrysogaster solstitialis* larva (Maibach and Goeldlin de Tiefenau, 1994).

Key

1.a. Base of wing yellow, with yellow veins (in contrast to grey veins in the tip 1/2). Male: face inflated, with ill-defined facial knob and mouth margin (figure 238) **>** 2

1.b. Base of wing not yellow, veins brown. Male: face not inflated **>** 3

2.a. Thoracic pleurae: proepimeron (between coxa 1 and anterior anepisternum) dusted, grey, strongly contrasting with other parts; face broad, in front view the distance between eyes equal to or wider then the width of an eye (figure 239, figure 240). Larger species: 6-8 mm (figure 241). Europe to Eastern Siberia **>** *Chrysogaster coemiteriorum* Linneaeus (= *C. chalybeata* Meigen)
Jizz: black species with bright red eyes and yellow wing bases, making the body appear contracted in the middle.

2.b. Thoracic pleurae: proepimeron shiny, equal to other parts; face smaller: in front view smaller than the width of an eye. Smaller species: 5-7 mm. Central Europe **>** *Chrysogaster basalis* Loew (probably = *Chrysogaster musatovi* Stackelberg, according to Speight (2003)

3.a. Face: in front view the distance between the eyes about twice the width of an eye. Male: scutellum with long hairs, at least on hind margin **>** 4

3.b. Face: in front view the distance between the eyes at most equal to the width of an eye (figure 242, figure 243). Male: scutellum almost bare, thoracic dorsum with very short hairs. 7-8 mm. Europe and North Africa **>** *Chrysogaster solstitialis* Fallén
Jizz: purple-black species with bright red eyes.

239.

broad

240.
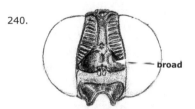
broad

241.

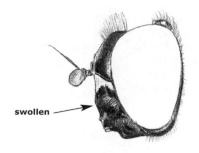
swollen

figure 238. *Chrysogaster cemiteriorum*, head of male (Verlinden).

figure 239, *Chrysogaster coemiteriorum*, face of male, front view.
figure 240. *Chrysogaster coemiteriorum*, face of female, front view.
figure 241. *Chrysogaster coemiteriorum*, habitus of male (Verlinden).

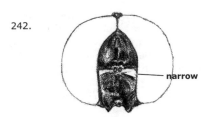

242.

narrow

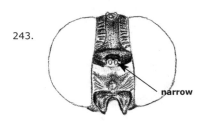

243.

narrow

figure 242. *Chrysogaster solstitialis*, face of male, front view.
figure 243. *Chrysogaster solstitialis*, face of female, front view (Verlinden).

4.a. Male: thoracic dorsum and scutellum covered with long erect, black hairs; facial knob indistinct and broad, almost absent. Female: thoracic dorsum covered with short light hairs; light hairs on the margin of the scutellum as well as covering all of its surface; hairs under the abdominal sternite 1 equal to width of femur 3 and equal to 1/2 this length on other sternites. 7-8 mm. Central Europe ❯ *Chrysogaster rondanii* Maibach and Goeldlin
Jizz: male with purple-reddish brilliance, female green.

4.b. Male: thoracic dorsum covered with semi–erect, light hairs; facial knob slightly more prominent and distinct. Female: hairs on thoracic dorsum virtually absent; scutellum with some hairs on hind margin, otherwise naked; hairs under sternite 1 distinctly shorter than the width of femur 3, shorter than 1/2 this width on the other sternites. 7-8 mm. Atlantic part of Central Europe ❯ *Chrysogaster virescens* Loew
Jizz: green species, reminiscent of *Orthonevra*.

CHRYSOSYRPHUS

Introduction

Chrysosyrpus are a genus of arctic hover-flies. They have black bodies and females lack a facial knob. They appear similar to *Lejogaster* and *Melanogaster*, but have a broader body. The key is adapted from Kassebeer (1995).

Key

1.a. Males ❯ 2
1.b. Females ❯ 3

2.a. Frons strongly swollen; eyes: contact line between eyes long, longer than longest hairs on ocellular triangle; face: pale hairs along the eyes; scutellum mainly pale-haired, thoracic dorsum has pale hairs among the black hairs, or pale ones may be absent. 7-9 mm. Boreal parts of Palaearctic ❯ *Chrysosyrphus niger* Zetterstedt
Jizz: shinier than *C. nasuta*, hairs on tergites erect.

2.b. Frons not swollen; eyes: contact line short, as long as the longest hairs on the ocellular triangle; face: dark hairs along the eyes; scutellum and thoracic dorsum: dark-haired. 7-8 mm. Boreal parts of Palaearctic ❯ *Chrysosyrphus nasutus* Zetterstedt
Jizz: duller than *C. nigra*, hairs on tergites inclined.

3.a. Thoracic dorsum with pale, erect hairs, rarely some black hairs in between; frons: distinctly swollen just before lunulae, laterally with transverse grooves. See 2a. ❯ *Chrysosyrphus niger* Zetterstedt

3.b. Thoracic dorsum with inclined black hairs; frons not swollen, in side view almost flat, and without clear transverse grooves. See 2b. ❯ *Chrysosyrphus nasutus* Zetterstedt

CHRYSOTOXUM

Introduction

Chrysotoxum are wasp mimics, and often found in association with open woodlands. They appear near the ground in the vegetation (e.g. *C. arcuatum*), on flowers (e.g. *C. bicinctum*), on leaves of shrubs and trees (e.g. *C. cautum*), and high up in trees (e.g. *C. octomaculatum*). The precise feeding habits of the larvae of *Chrysotoxum* are unknown (Rotheray, 1993). The larvae live in the ground, associated with ant nests and are presumably predators of root aphids. There are reports of females ovipositing near ant nests. *C. bicinctum* larvae have been reared on pea aphids in the laboratory, supporting the predatory life history (Rotheray, 1993).

Recognition

Chrysotoxum contains broad-bodied, black and yellow hoverflies with long antennae. They mimic social wasps and Speight (2003) remarks that *C. fasciolatum* is a surprisingly good mimic of *Vespula* queens. The key is based on Van der Goot (1981), Sack (1932), Seguy (1961), Smit, Renema and Van Aartsen (2001), and Speight (2003). The names *C. festivum* and *C. arcuatum* have been preserved by the Zoological Code Commission.

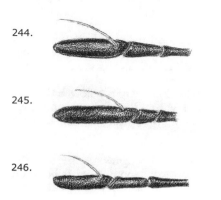

244.

245.

246.

figure 244. *Chrysotoxum cautum*, antenna.
figure 245. *Chrysotoxum intermedia*, antenna.
figure 246. *Chrysotoxum verralli*, antenna (Verlinden).

Key

1.a. Antennae: 3rd segment long: as long as or longer than segments 1 and 2 together (figure 244, figure 245) ❯ 2
1.b. Antennae: 3rd segment short: shorter than segments 1 and 2 together (figure 246) ❯ 6

2.a. Tergites 3 and 4 with short hairs, shorter than the height of metatars 1, seemingly bare ❯ 3
2.b. Abdominal tergites with long hairs, longer than the height of metatars 1 ❯ 4

3.a. Frons yellow; abdominal margin of tergites mainly yellow; antennae: segment 3 equal in length to the sum of segments 1 and 2. 10-14 mm. Southern part of Central Europe ❯ **Chrysotoxum cisalpinum** Rondani
3.b. Frons black; abdominal margin black and yellow; antennae: segment 3 distinctly longer than segments 1 and 2 together. 10-14 mm. Central and Southern Europe, North Africa, east to Afganistan ❯ **Chrysotoxum intermedium** Meigen
Jizz: blackish *Chrysotoxum*, yellow bands run from hind corner of side margin to the middle of the front margin.
Note: probably a species complex with more species in Southern Europe.

4.a. Antennae: 3rd segment longer than segments 1 and 2 together (as figure 245); abdomen strongly arcuate; male: genitalia small, at most reaching the hind margin of sternite 4 (figure 247) ❯ 5
4.b. Antennae: 3rd segment as long as segments 1 and 2 together (figure 244); abdomen relatively flat (use reference material). Male: genitalia very large, reaching over the hind margin of sternite 4 (figure 248). Female: unique among European *Chrysotoxum* in possessing a longitudinal, median, membranous strip on abdominal tergite 6, which divides this tergite into 2 parts; habitus: figure 249. 12-14 mm. Europe, east to Western Siberia ❯ **Chrysotoxum cautum** Harris
Jizz: large, yellowish *Chrysotoxum*, often suns itself on or patrols along leaves, wing without well-marked dark spot.

247.

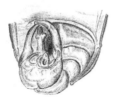

248.

figure 247. *Chrysotoxum fasciolatum*, tip of abdomen from below.
figure 248. *Chrysotoxum cautum*, tip of abdomen from below (Verlinden).

5.a. Wing: front margin hyaline, except for the stigma, without dark spot in the top 1/2; scutellum yellow with a darkened centre; abdomen short and oval, less than 1.5 times as long as wide, in side view thick (figure 250). Smaller species: 9-12 mm. Northern Europe and elevated parts of Central Europe, east to Japan ❯ *Chrysotoxum arcuatum* Linneaus (= *Chrysotoxum fasciatum* auctorum)
Jizz: rather small, with equal coverage of black and yellow at first glance. Often foraging and flying close to the ground.

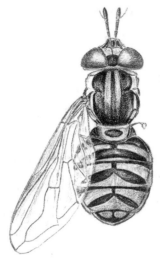

figure 249. *Chrysotoxum cautum*, habitus (Verlinden).

250.

251.

figure 250. *Chrysotoxum arcuatum*, abdomen, side view.
figure 251. *Chrysotoxum fasciolatum*, wing (Verlinden).

5.b. Wing: front margin yellowish with a distinct dark spot in the top 1/2 (figure 251); scutellum yellow with a broad dark hind margin; abdomen elongate, about 1.5 times as long as wide, and not that thick. Larger species: 13-17 mm. Northern and Central Europe, east to Japan, Nearctic ❯ *Chrysotoxum fasciolatum* Degeer

6.a. Abdominal tergites 2-5 have yellow, usually interrupted bands of equal width ❯ 7
6.b. Abdomen with distinct yellow bands on tergites 2 and 4 only (figure 252), the other tergites black (typical specimens), or with thin yellow bands (var. *tricinctum*). 10-11 mm. Europe, North Africa, east to Siberia ❯ *Chrysotoxum bicinctum* Linneaus
Jizz: Two strong, uninterrupted, yellow bands. In flight similar to *D. tricinctus* (short antennae).

7.a. Abdominal margin black, the yellow bands do not reach the margin (figure 253); wing: dark spot at the front in the top 1/2 ❯ 8
7.b. Abdominal margin yellow and black, the yellow bands reach the margin (figure 254); wing with or without dark spot ❯ 10

8.a Antennae: 1st segment longer than the 2nd segment; femora 2 and 3: black on basal half; wing: dark spot large and elongate; abdomen elongate with parallel margins; scutellum: black with small yellow hind margin (figure 255). 10-12 mm. Central Europe, east to Kazakstan ❯ *Chrysotoxum lineare* Zetterstedt
Jizz: abdomen elongate, a dark species.

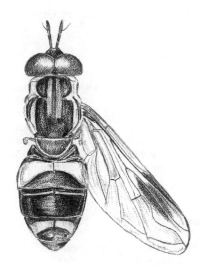

figure 252. *Chrysotoxum bicinctum*, habitus of male (Verlinden).

yellow line

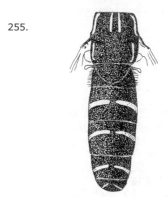

8.b. Antennae: 1st and 2nd segments of equal length; femora 2 and 3: entirely yellow or black on basal 1/4; wing: dark spot small, square; abdomen oval; scutellum: black with all broad yellow margins or almost entirely yellow **❯** 9

9.a. Femora 1 and 2 completely yellow; katepisternum normally with yellow spot. 12-15 mm. Europe, North Africa, east to Japan **❯** ***Chrysotoxum festivum*** Linneaus (= *Chrysotoxum arcuatum* auctorum)

Jizz: blackish *Chrysotoxum* with well-defined yellow bands, similar to next.

9.b. Femora 1 and 2 yellow with black bases; katepisternum normally without yellow spot. 10-13 mm. Northern and Central Europe, east to Siberia **❯** ***Chrysotoxum vernale*** Loew

Jizz: blackish *Chrysotoxum* with well-defined yellow bands, similar to previous.

10.a. Tergites 3 and 4: black front margins not interrupted, front margin continuously black (figure 256, figure 257); scutellum yellow- or black-haired; antennal knob in dorsal view: rectangular, protruding, undusted part broad, about 1.75 or more times as wide as long; female: hairs on side margins of thorax and abdomen, but especially on the postalar lobes, long and fine **❯** 11

figure 253. *Chrysotoxum festivum*, abdomen (Verlinden).
figure 254. *Chrysotoxum octomaculatum*, abdomen (Verlinden).
figure 255. *Chrysotoxum lineare*, thorax and abdomen (Sack, 1930).

10.b Tergites 3 and/or 4: black front margins interrupted at the side margin by a yellow stripe, the side margin itself black again (figure 254) (although variable, most specimens exhibit this feature on at least 1 of the tergites); scutellum yellow-haired; antennal knob in dorsal view: the rectangular, protruding, undusted part elongate, about 1.25-1.5 as wide as long; female: hairs on side margins of thorax and abdomen, but especially on the postalar lobes, very short. 10-13 mm. Central and Southern Europe, east to Kazakstan **❯** ***Chrysotoxum octomaculatum*** Curtis

Jizz: very similar to *C. verralli*, but more rounded (a small *C. cautum*) and scutellum yellow-haired.

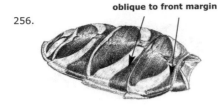

256.

oblique to front margin

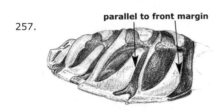

257.

parallel to front margin

figure 256. *Chrysotoxum elegans*, abdomen.
figure 257. *Chrysotoxum verralli*, abdomen
(Verlinden).

11.a. Tergite 2: the black front margin increases in width towards the side margin because the front margin of the yellow band does not follow the tergite front margin (figure 256); scutellum yellow-haired ❯ 12

11.b. Tergite 2: the black front margin of equal width over much of its length because the front margin of the interrupted, yellow band follows the tergite front margin closely and separates only at the side margin (figure 257); scutellum black-haired. 10-13 mm. Central Europe, east to Siberia ❯ *Chrysotoxum verralli* Collin
Jizz: yellow *Chrysotoxum*, similar to *C. octomaculatum*, but more elongate. Yellow spots on tergite 2 appear triangular.

12.a. Tergites 3 and 4: yellow hind margins small, smaller than the yellow band, or lacking. 10-13 mm. Europe, east to Caucasus ❯ *Chrysotoxum elegans* Loew
Jizz: blackish *Chrysotoxum*. Yellow spots on tergite 2 appear as stripes.

12.b. Tergites 3 and 4: yellow hind margins broad, as broad as the yellow band. 10-13 mm. Europe ❯ *Chrysotoxum latilimbatum* Collin
Note: most probably only a colour variation of the former and listed as synonym by Peck (1988).

CRIORHINA

Introduction

Criorhina are woodland species and mainly visit the flowers of trees and shrubs. I found numerous *C. pachymera* visiting *Acer campestre* flowers. Many species maintain territories at the bases of tree trunks or seem to inspect the bark, travelling from tree to tree. Their larvae live in dead wood.

Recognition

Criorhina are large bee and bumblebee mimics. They can be separated from other large bee and bumblebee mimics (e.g. *Mallota, Arctophila, Pocota, Brachypalpus*) by the typical form of their antennae: the third segment is (much) shorter than it is wide, while the first segments are much thinner and form a stalk. In *Criorhina*, the face projects downwards, in contrast to *Pocota* and *Brachypalpus*. *Criorhina* species fall into two groups:
• Bumblebee mimics: *C. berberina, C. ranunculi* and *C. floccosa*.
• Honeybee mimics: *C. pachymera* and *C. asilica*.

Key

1.a. Femur 3: strongly thickened and curved, especially in the male (figure 258, figure 259, figure 260, figure 261); tibia 1 and 2 with long, erect hairs at the back ❯ 2

1.b. Femur 3: marginally thickened and not curved; tibia 1 and 2 with short, adpressed hairs at the back ❯ 3

2.a. Thorax and abdomen with long, dense hairs, a bumblebee mimic; hairs black on thorax and front part of abdomen, reddish on tip of abdomen (sometimes scutellum with pale hairs), but in rare cases thorax and abdomen with greyish hairs; thoracic dorsum without stripes of pale dust. 14-17 mm. Northern and Central Europe ❯ *Criorhina ranunculi* Panzer

258.

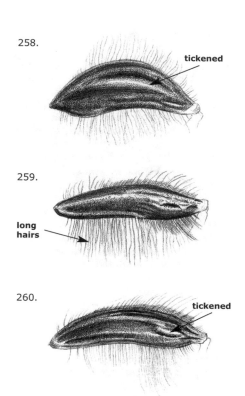

tickened

259.

long
hairs

260.

tickened

261.

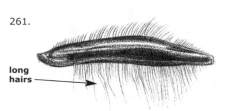

long
hairs

figure 258. *Criorhina pachymera*, femur 3 of male.
figure 259. *Criorhina pachymera*, femur 3 of female.
figure 260. *Criorhina ranunculi*, femur 3 of male.
figure 261. *Criorhina ranunculi*, femur 3 of female (Verlinden).

2.b. Thorax and abdomen with long, but less dense hairs, a honeybee mimic; thorax and abdomen with pale brown hairs, tip of abdomen may be black-haired; thoracic dorsum with weak stripes of pale dust. 13-17 mm. Central Europe ❯ *Criorhina pachymera* Egger

3.a. Tergites 2-4 without grey or yellowish dusted bars; bumblebee mimics ❯ 4

figure 262. *Criorhina asilica*, habitus of male (Verlinden).

3.b. Tergites 2-4 with grey or yellowish dust bars and loose, yellowish hairs (figure 262), tergite 2 without distinct side tufts of yellow hairs; a honeybee mimic. 12-14 mm. Northern and Central Europe ❯ *Criorhina asilica* Fallén

4.a. Tergite 2 with distinct side tufts of long yellow hairs, tergites 3 and 4 with short, dense reddish pile and dust, only black at side margins; tibia 1 and 2 with pale hairs, short and adpressed (figure 263). 12-13 mm. Central Europe ❯ *Criorhina floccosa* Meigen (= *Brachymyia floccosa* Meigen)

4.b. Tergite 2 without distinct side tufts of yellow hairs, tergites 3 and 4 without short dense reddish pile; thorax and abdomen covered with long, dense hairs in 2 colour forms: a completely yellowish-orange form (var. *oxycanthae*) and a form with the thoracic dorsum reddish-haired, the scutellum and basal part of abdomen black-haired, and the tip of the abdomen whitish-haired (typical form, figure 265); tibia 1 and 2 with black hairs, short and adpressed (figure 264). 8-13 mm. Central Europe ❯ *Criorhina berberina* Fabricius (= *Brachymyia berberina* Fabricius)

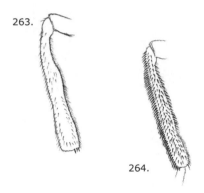

figure 263. *Criorhina floccosa*, tibia 1.
figure 264. *Criorhina berberina*, tibia 1
(Verlinden).

figure 265. *Criorhina berberina*, habitus of
male (Verlinden).

DASYSYRPHUS

Introduction

Dasysyrphus dwell in forests, along forest
edges and probably also occur high in
trees. They visit all kind of flowers with
shallow nectaries. Many species are shade
tolerant and can be found inside the forest. Males hover under tree cover. Females
wander through the forest, sometimes
deep in the vegetation. Their larvae prey
on aphids.

Recognition

Dasysyrphus have yellow on black, oval
bodies. They are similar to many Syrphini
genera, in particular *Eupeodes* and
Parasyrphus, but their eyes are haired
(*Eupeodes* and most *Parasyrphus* have bare
eyes) and the abdomen is spotted rather
than banded (except for *D. tricinctus* and
D. eggeri). The key is based on Van der
Goot (1981), Verlinden (1991), Bradescu
(1991), Doczkal (1996) ,Reemer (2002)
and Bartsch et al. (2009a). As Speight
(2003) points out, *D. pauxillus, D. pinastri,
D. lenensis* and *D. nigricornis* are not well-
defined at the moment. *D. postclaviger* and
D. nigricornis are added after the interpretation of Bartsch et al. (2009a).

Key

1.a. Tergites 3 and 4 with equally-sized
yellow patterns; tergite 2: pale pattern
always present, equally broad or broader
than pattern on tergites 3 and 4 **❯ 2**
1.b. Tergite 3 with yellow pattern much
broader than on tergite 4; yellow pattern
on tergite 2 small or absent (figure 266).
10-12 mm. Northern and Central
Europe, east to Japan **❯ *Dasysyrphus
tricinctus* Fallén**
Jizz: broad, blackish species with small yellow
bands on the abdomen, in flight quite similar to
flying *Chrysotoxum bicinctum*.

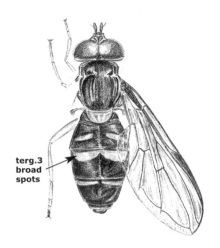

terg.3
broad
spots

figure 266. *Dasysyrphus tricinctus*, habitus of
male (Verlinden).

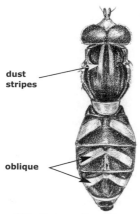

dust
stripes

oblique

figure 267. *Dasysyrphus albostriatus*, habitus of male (Verlinden).

2.a. Thoracic dorsum with 2 or 4 longitudinal stripes of white dust, tergites with spots (which may broadly connect) at a distinct angle to the front margin (figure 267) ❯ 3

2.b. Thoracic dorsum without stripes of white dust; tergites with well separated spots parallel to the front margin (or at most at a small angle) ❯ 4

3.a. Thoracic dorsum with 2 stripes of white dust on the front 1/2 of the dorsum; tergites with linear spots, usually separated, but they may connect on tergites 3 and 4. 8-10 mm. Europe to Japan ❯ **Dasysyrphus albostriatus** Fallén
Jizz: oblique yellow spots on abdomen, 2 white stripes on thorax and black pterostigma.

3.b. Thoracic dorsum with 4 stripes of white dust over the full length of the dorsum, the inner 2 most densely marked; tergites with oval spots that broadly connect on tergites 3 and 4. 8-10 mm. Eastern part of Central Europe, Pyrenees, east to Mongolia ❯ **Dasysyrphus eggeri** Schiner
Jizz: broad yellow spots or bands on abdomen, 4 white stripes on thorax.

4.a. Tergites 3 and 4: lateral margin black and yellow: yellow spots reach the margin; antennae pale or dark ❯ 5

4.b. Tergites 3 and 4: lateral margin black, the yellow spots do not reach it; antennae always dark, at most 3rd segment turning pale below ❯ 8

5.a. Tergites 3 and 4: spots strongly narrowed to less than 0.5 times their largest width (figure 269); antennae pale, dark-brown or black ❯ 6

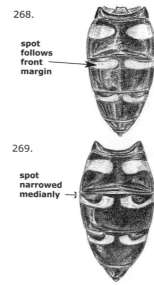

268.

spot
follows
front
margin

269.

spot
narrowed
medianly →

figure 268. *Dasysyrphus hilaris*, abdomen of female.

figure 269. *Dasysyrphus friuliensis*, abdomen of female (Verlinden).

5.b. Tergites 3 and 4: spots hardly narrowed, at most to 0.6 times their largest width (figure 268); antennae pale ❯ 7

6.a. Antennae pale; scutellum: hind margin black, scutellar dorsum predominantly yellow haired; face with pale hairs. 7-11 mm. Northern, Central and Southeastern Europe.❯ **Dasysyrphus postclaviger** Stys and Moucha

6.b. Antennae brown to black; scutellum: hind margin yellow, scutellar dorsum predominantly black haired; face predominantly with black hairs. 10-12mm. Northern, Central and Southeastern Europe. ❯ **Dasysyrphus friuliensis** Van der Goot.

7.a. Face with black longitudinal stripe (figure 270). Female: sternite 2 black at hind margin; frons: dust spots faintly marked. 8-10 mm. Northern and Central Europe, Nearctic ❯ **Dasysyrphus venustus** Meigen
Note: Bartsch et al. (2009a) remarks that *D. venustus* north of the polar circle can be quite dark and small, resembling *D. pauxillus* and *D. nigricornis*

7.b. Face yellow, at most mouth edge and facial knob darkened (figure 271). Female: sternite 2 pale; frons: sharply marked dust spots. 10-12 mm. Northern and Central Europe ❯ **Dasysyrphus hilaris** Zettersted
Note: *D. venustus* and *D. hilaris* may be found to hide a species complex in the future.

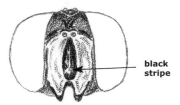

figure 270. *Dasysyrphus venustus*, face of female (Verlinden).

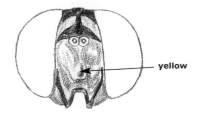

figure 271. *Dasysyrphus hilaris*, face of female (Verlinden).

8.a. Scutellar dorsum predominantly yellow haired; lower part of face black haired or with a mixture of black and pale hairs. Male: hairs on the side margin of tergite 2 black; angle of approximation of eyes blunt or acute. Female: dust patches on frons small, only present around eye margin and when extending into the middle, their width in the middle is less than 1/5th of the distance between lunulae and ocelli ❭ 9

8.b. Scutellar dorsum predominantly black haired; lower part of face pale haired. Male: hairs on the side margin of tergite 2 yellowish (with black ones in hind corner); angle of approximation of eyes acute. Female: dust patches on frons large and (almost) touching, their width in the middle is 1/4th of the distance between lunulae and ocelli or wider ❭ 10

9.a. Tibia and tars 3: mainly pale, with some diffuse black pattern; sternite 2: yellow with blackish crossband (figure 274); habitus as figure 272 and figure

273. Males: angle of approximation of eyes blunt, more than 90 degree; eyes meeting over a distance equal to the frons length. Female: dust patches on frons small, only present around eye margin, although they extend into the middle, they hardly touch. 5-8mm. Northern and elevated parts of Central Europe. ❭ ***Dasysyrphus pauxillus*** Williston

Jizz: small, dark Dasysyrphus with relatively pale legs.

9.b. Tibia and tars 3: blackish with the base of the tibia pale. Male: angle of approximation of eyes acute, less than 90 degree; eyes meeting over a distance shorter than de frons length. Female: dust patches extending into the middle forming a band. 5-8mm. Northern Europe above polar circle. ❭ ***Dasysyrphus nigricornis*** Verral

Jizz: small, dark Dasysyrphus with very dark legs.

figure 272. *Dasysyrphus pauxilus*, habitus of female (Verlinden).

figure 273. *Dasysyrphus pinastri*, habitus of female (Verlinden).

274.

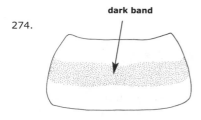

dark band

275.

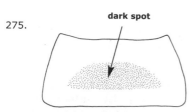

dark spot

276.

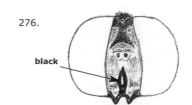

black

277.

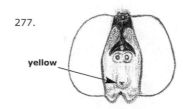

yellow

figure 274. *Dasysyrphus pauxilus*, sternite 2.
figure 275. *Dasysyrphus pinastri*, sternite 2
(Doczkal, 1996).

figure 276. *Didea intermedia*, face of female.
figure 277. *Didea fasciata*, face of female
(Verlinden).

10.a. Sternite 2: yellow with blackish, oval patch (figure 275). Male: tars 1 with yellow bristles. Female: frons with a band of dust, its width in the middle is 1/3 the distance between front ocellus and lunula Habitus fig. 273. 7-11 mm. Northern Europe and elevated parts of Central Europe, east to Siberia ❯ *Dasysyrphus pinastri* Degeer

10.b. Sternite 2: yellow with blackish crossband (as figure 274). Female: frons with large patches of dust, which may form a band, with a width in the middle of 1/4 the distance between front ocellus and lunula. 9-10 mm. Central Europe, Eastern Siberia ❯ *Dasysyrphus lenensis* Bagatshanova

DIDEA

Introduction

Didea can be found on the edges of nutrient-poor deciduous and mixed forests. *D. fasciata* and *D. intermedia* migrate over large distances and occur everywhere in autumn. They are large hoverflies with characteristic 'ski-glasses'-shaped bands on their broad, flat abdomens. Their larvae prey on aphids.

Key

1.a. Halteres with black knob; face yellow with dark facial knob and mouth edge, usually joined in a stripe (figure 276) ❯ 2
1.b. Halteres with yellow knob; face yellow, at most the tip of the facial knob darkened (figure 277); habitus as figure 278. 11-13 mm. Europe, east to North India, Nearctic ❯ *Didea fasciata* Macquart
Jizz: abdomen broad and flat. In *Didea* with the broadest bands, tergites 2 and 3 often more than 1/2 yellow.

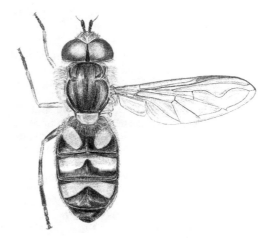

figure 278. *Didea fasciata*, habitus of male
(Verlinden).

279.

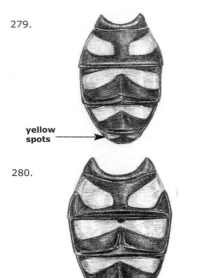

yellow
spots →

280.

black →

figure 279. *Didea intermedia*, abdomen of female.
figure 280. *Didea alneti*, face of female (Verlinden).

figure 281. *Didea alneti*, face of female (Verlinden).

2.a. Tergite 5 with yellow spots or largely yellow (figure 279). Male: tergite 4 with front margin of wedge-shaped bars mostly flush with front edge of tergite. Female: tergite 5 with hair on lateral margin black. Smaller: 7-12 mm. Europe, east to Kamchatka **> *Didea intermedia* Loew**
  Jizz: abdomen broad, yellow bands smaller than in *fasciata* and tergites 2 and 3 less than 1/2 yellow.
2.b. Tergite 5 black, rarely with spots (figure 280). Male: tergite 4 with front margin of wedge-shaped bars clearly separated from the front margin of the tergite or only touching front of tergite about centre line. Female: tergite 5 with long white hairs in addition to fringe of black hairs; face as figure 281. Larger: 12-16 mm. Northern Europe and elevated parts of Central Europe, Nearctic **> *Didea alneti* Fallén**
  Jizz: large species, broad abdomen with typical translucent, greenish spots in live specimens.

DOROS

Introduction

Large hoverflies with a petiolate body, that live in oak-birch forests often mixed with pines. The adults have a very limited flight period of one to two weeks. Finding this species requires being in the right place at the right moment! The larvae are associated with ant nests.

Key

1. Large hoverfly with strong petiolate abdomen and darkened front margin of wing (figure 282). Abdomen with 3 yellow bands, thorax black with yellow bands at the side margin. 14-16 mm. Central and Southern Europe **> *Doros profuges*** Harris (= *Doros conopseus* Linnaeus)

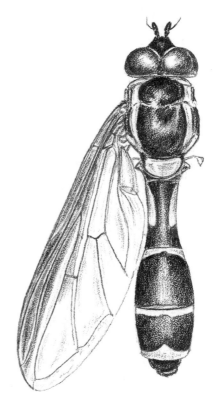

figure 282. *Doros profuges*, habitus of male (Verlinden).

EPISTROPHE

Introduction

Epistrophe are associated with forests, where they dwell in open spots and forest edges and are found on flowers and sunlit leaves. They probably also occur in the canopy layer. *E. eligans* and *E. ochrostoma* appear in March to April. *E. grossulariae* is among the last to occur; the adults can be found on flowers in forest edges and grasslands as late as September. The larvae are aphid predators, and are usually associated with aphids on trees, but also occur on shrubs and tall herbs (Rotheray, 1993). *E. flava* larvae have been collected from aphid galls on *Malus* (see (Speight, 2003)). *Epistrophe* larvae have a characteristic flattened shape and green colour, a good camouflage on tree leaves. Larvae of *E. grossulariae* seem to live for one to three years before pupating (Rotheray, 1993).

Recognition

Most species of *Epistrophe* are broad-bodied, large hoverflies with large, rectangular yellow bands on a black abdomen. They

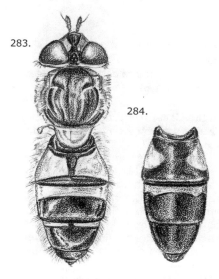

283.

284.

figure 283. *Epistrophe eligans*, habitus of female.
figure 284. *Epistrophe eligans*, abdomen (Verlinden).

may be mistaken for a *Syrphus*, but *Epistrophe* has a yellow stripe on the margins of the thoracic dorsum and many species are stouter. The banded *Parasyrphus*-species are smaller. The key is based on Van der Goot (1981), Doczkal and Schmid (1994) and Bartsch et al. (2009a).

Key

1.a. Tergite 4 with spots or completely black, tergite 3 with spots or bands ❯ 2
1.b. Tergites 3 and 4 with bands ❯ 4

2.a. Tergites 3 and 4 with equally large spots, tergite 4 never black ❯ 3
2.b. Tergite 4 with smaller yellow markings than tergite 3 or tergite 4 black (figure 283, figure 284). 10-11 mm. Central and Southern Europe, east to the Caucasus ❯ *Epistrophe eligans* Harris

3.a. Tergites 3 and 4: with broadly separated, oblique, undusted spots (rarely connected at front margin), spots on tergite 2 smaller than on tergites 3 and 4; tergites 2-4 lateral margin often with small yellow spots at the front corner; tergite 5 with yellow markings (figure 285, figure 286); antennae pale. 8-10 mm. Europe, east to Siberia ❯ *Epistrophe (Epistrophella) euchroma* Kowarz
   Jizz: slender species, reminiscent of *Melangyna*, but with large, square spots with small spots in front of them.
3.b. Tergites 3 and 4: the interrupted bands heavily dusted, the yellow colour hardly visible and sometimes even lacking; tergite 5 black; tergites 3-8: lateral margin black; antennae black. 11-12 mm. Elevated parts of Central Europe, mountains ❯ *Epistrophe leiophtalma* Schiner & Egger

4.a. Antennae: entirely or mainly black; thoracic dorsum dull; frons entirely or partly with yellow hairs; femora 1 and 2: all long hairs yellow ❯ 5
4.b. Antennae: yellow, upper margin of 3rd segment often darkened (in case the 3rd segment is largely black, then the thoracic dorsum shiny or frons entirely with black hairs or femora 1 and 2 with long black hairs) ❯ 7

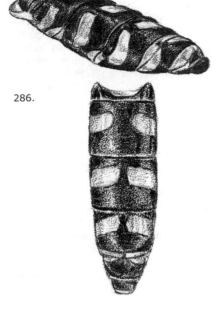

285.

286.

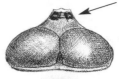

287.

288.

figure 287. *Epistrophe grossulariae*, head of male.
figure 288. *Epistrophe grossulariae*, head of female (Verlinden).

289.

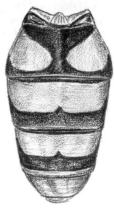

290.

figure 285. *Epistrophe euchroma*, abdomen of male.
figure 286. *Epistrophe euchroma*, abdomen of female (Verlinden).

5.a. Mouth edge yellow; tergites 3 and 4: yellow bands not hidden by dust **>** 6
5.b. Mouth edge partly black; tergites 3 and 4: the bands heavily dusted, the yellow colour scarcely visible and sometimes absent. Specimens with connected spots on tergite 4, see 3 **>** *Epistrophe leiophtalma* Schiner & Egger

6.a. Frons: black above lunula (figure 287, figure 288), covered with black hairs for more than 1/2 its area; scutellum with black hairs posteriorly; femora: at least the bases of femora 1 and 2 and often those of femur 3 black on basal 1/4; tergite 5 with black band. 12-14 mm. Europe, in Asia east to Kamchatka, Nearctic **>** *Epistrophe grossulariae* Meigen
Jizz: yellow bands on tergites 3 and 4 straight, neither curved nor narrowed.
6.b. Frons: yellow above lunula (figure 289), covered entirely or largely with yellow hairs; scutellum with yellow hairs only; femora completely yellow; tergite 5 yellow (figure 290). 10-11 mm. Northern and Central Europe, in Asia east to the Pacific coast **>** *Epistrophe diaphana* Zetterstedt

figure 289. *Epistrophe diaphana*, head of female.
figure 290. *Epistrophe diaphana*, abdomen of female (Verlinden).

7.a. Tars 3: entirely pale or dark, without distinct dark rings on pale background – **>** 8
7.b. Tars 3: background colour pale, with 6 well defined black rings, on the base and top of the metatars and the apex of the other tarsal segments; scutellar dorsum mainly black haired; arista black; tibia 3 pale; tergite 5 yellow with a small black spot in the middle. 11-13mm. Finland. **>** *Epistrophe annulitarsis* (Stackelberg).
Jizz: similar to E. flava, tars 3 quite different.

8.a. Wing: basal cells bm and/or br partly bare **>** 9
8.b. Wing: basal cells bm and br entirely covered by microtrichia **>** 11

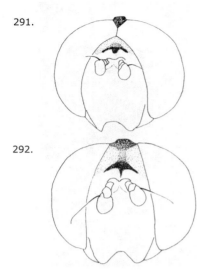

291.

292.

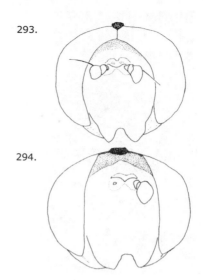

293.

294.

figure 291. *Epistrophe flava*, face of male.
figure 292. E*pistrophe flava*, face of female
(Doczkal and Schmid, 1994).

figure 293. *Epistrophe ochrostoma*, face of
male.
figure 294. *Epistrophe ochrostoma*, face of
female (Doczkal and Schmid, 1994).

9.a. Thoracic dorsum: in front 2/3rth (almost) without dusting, shiny; scutellum: at least with 1/3rth black hairs; tergite 5: black spot at least 1/3rth of the tergite width ❭ 10

9.b. Thoracic dorsum: heavily dusted, dull, with 2 clear stripes of white pollinosity in front; scutellum yellow-haired; tergite 5 yellow, at most with a small black spot, at most 1/5th of the tergite width; arista black. Male: face as figure 291. Female: frons at and above depression dusted (figure 292). 11-13 mm. Northern and Central Europe, in Asia east to Pacific coast ❭ **Epistrophe flava** Doczkal and Schmid

Jizz: stout species, with greenish dull thoracic dorsum with dust stripes and broad yellow bands on tergites.

10.a. Arista black; frons dusted only along eye margin, shiny (or nearly so) in the middle. 11-12 mm. Northern and Central Europe, in Asia east to Pacific coast ❭ **Epistrophe nitidicollis** Meigen

Jizz: thoracic dorsum blackish and shiny and tergites equally black and yellow. Easily confused with E. melanostoma, but slightly narrower tergite 5 with black band and scutellum black-haired.

10.b. Arista yellow; frons evenly dusted in posterior half, gradually less so towards the lunulae. 9-12 mm. Sweden, Finland, Siberia. ❭ **Epistrophe olgae** Mutin

11.a. Frons partly yellow; tibia 3 pale without dark ring; tergite 5 yellow, at most with small triangular black spot in middle ❭ 12

11.b. Frons black; tibia 3 pale with poorly marked dark ring; tergite 5 with broad black band ❭ 14

12.a. Face in front view smaller (less than 1/2 as broad as high); mouth edge black or yellow; mouth opening larger, distance between mouth edge and eye margin small (less than 0.6 mm); clypeus (inside mouth opening, at the front) broad and medianly dusted. Males: eyes: angle of approximation 85-102° ❭ 13

12.b. Face in front view very broad (more than 1/2 as broad as high); mouth edge yellow; mouth opening small, distance between mouth edge and eye margin long (more than 0.6 mm); clypeus (inside mouth opening, at the front) narrow and entirely shiny; frons above lunulae yellow, in the female with a sharp border between yellow and black above it (figure 293, figure 294). Male: eyes: angle of approximation 104-110°. 10-12 mm Probably Northern and Central Europe ❭ **Epistrophe ochrostoma** Zetterstedt

Jizz: broad, yellow head.

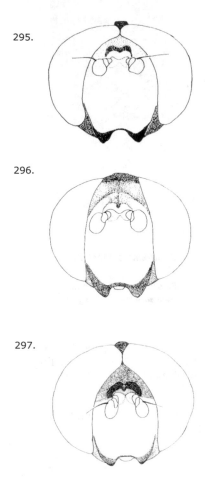

295.

296.

297.

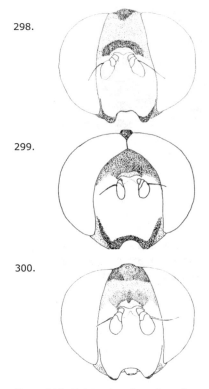

298.

299.

300.

figure 298. *Epistrophe obscuripes*, face of female.
figure 299. *Epistrophe cryptica*, face of male.
figure 300. *Epistrophe cryptica*, face of female (Doczkal and Schmid, 1994).

figure 295. *Epistrophe melanostoma*, face of male.
figure 296. *Epistrophe melanostoma*, face of female.
figure 297. *Epistrophe obscuripes*, face of male (Doczkal and Schmid, 1994).

13.a. Mouth edge yellow (figure 291); thoracic dorsum: heavily dusted, dull, with 2 clear stripes of white pollinosity in front; tars 3 darkened. See 9 **>**
Epistrophe flava Doczkal and Schmid
13.b. Mouth edge largely black (figure 295, figure 296); thoracic dorsum: in front 2/3 (almost) without dusting, shiny; tars 3 partly pale. 10-12 mm. Northern and Central Europe **>**
Epistrophe melanostoma Zetterstedt
Jizz: easily confused with *E. nitidicollis*, but slightly broader, tergite 5 mainly yellow and scutellum yellow-haired.

14.a. Face in front view smaller, less than 1/2 as broad as high; genae and often lateral parts of face with many black hairs; frons: a large undusted part present above lunulae (figure 297, figure 298); arista: black, contrasting with pale 3rd antennal segment. Female: mouth edge black (figure 298). 10-12 mm. Confirmed from continental parts of Central Europe **>**
Epistrophe obscuripes Strobl (= *Epistrophe similis* Doczkal and Schmid)
14.b. Face in front view broader, more than 1/2 as broad as high; genae and lower, lateral parts of face without black hairs; frons: entirely dusted, at most with a small bare spot (figure 299, figure 300); arista: pale, same colour as 3rd antennal segment. Female: mouth edge almost entirely yellow (figure 300). 10-12 mm. Confirmed from continental parts of Central Europe and Sweden **>**
Epistrophe cryptica Doczkal and Schmid

EPISYRPHUS

Introduction

Episyrphus contains the familiar *E. balteatus*, one of the most common hoverflies and a regular visitor to gardens and houses. The other *Episyrphus* species are common forest and shrub species. They are also found in gardens with relatively mature shrubs and trees, but are very variable in abundance. Their larvae prey on aphids. The genus *Meliscaeva* is included as a subgenus of *Episyrphus*, but later works reinstall *Meliscaeva* as separate genus.

Recognition

All species are medium-sized with an elongated body and bare eyes. *E. balteatus* is unique in its rather orange body colour and double-banded tergites. *E. auricollis* and *E. cinctella* have yellow bands or spots and may be confused with *Meligramma* or *Melangyna*. *E. cinctella* is similar to *Melangyna cincta*, but the yellow spots on tergite 2 are square instead of triangular in *E. cinctella*.

Key

1.a. Tergites 3 and 4: black with yellow to orange spots or band; thoracic dorsum without light stripes **>** 2
1.b. Tergites 3 and 4 (figure 301): orange with a black 'moustache' in the front half and a black hind border; shiny greenish thorax with pale longitudinal stripes. 9-12 mm. Cosmopolitic **>** *Episyrphus balteatus* Degeer
Jizz: orange abdomen, each tergite has narrow double bands (the upper often divided like a moustache). When it is cold during pupation, the black on the abdomen becomes extensive.

2.a. Face: lunula yellow in contrast to black spot above it (figure 302), at least facial knob black; wing: alula triangular, widening from base to tip; tergite 2 usually with small elongate yellow spots, or small triangular spots; tergites 3 and 4 with a pair of spots (form *maculicornis*, figure 304, figure 305), which may extend into a band (typical form, figure 306).

figure 301. *Episyrphus balteatus*, habitus of male (Verlinden).

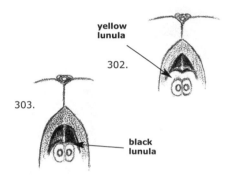

figure 302. *Episyrphus auricollis*, frons.
figure 303. *Episyrphus cinctellus*, frons (Verlinden).

9-11 mm. Europe, North Africa **>** *Episyrphus (Meliscaeva) auricollis* Meigen
Jizz: elongated body, tergites 3 and 4 with spots or narrowly connected bands that usually do not reach the abdominal margin.

2.b. Face: lunula black, not contrasting with black spot above it (figure 303), face entirely yellow; wing: alula narrow, rectangular, with parallel front and hind margin; tergite 2 with broad, square yellow spots; tergites 3 and 4 with broad yellow bands which usually reach the abdominal margin (figure 307, figure 308). 9-10 mm. Europe, North Africa, in Asia east to Pacific **>** *Episyrphus (Meliscaeva) cinctellus* Zetterstedt
Jizz: elongated body, tergites 3 and 4 with bands that in typical specimens have little points in the middle, one pointing forwards and the other backwards.

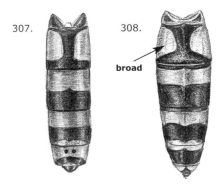

figure 304. *Episyrphus auricollis maculicornis*, habitus of male.
figure 305. *Episyrphus auricollis maculicornis*, abdomen of female.
figure 306. *Episyrphus auricollis auricollis*, abdomen of male (Verlinden).

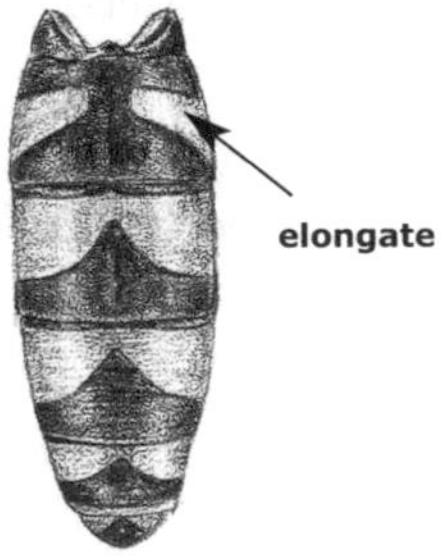

figure 307. *Episyrphus cinctellus*, abdomen of male.
figure 308. *Episyrphus cinctellus*, abdomen of female (Verlinden).

ERIOZONA

Introduction

Eriozona contains two quite different species: the bumblebee mimic *E. syrphoides* and the yellow banded *E. erraticus*. Both are forest species, with a preference for coniferous forests. The larvae of *E. syrphoides* have been found on *Picea* feeding on *Cinara* aphids (Kula, 1983). The generic position of *E. erratica* is under debate, some place it under *Megasyrphus*.

Recognition

E. syrphoides is a large bumblebee mimic, similar to *Eristalis oestracea, Volucella bombylans, Cheilosia illustrata* and *Mallota fuciformis* in the field. The latter two species have a black scutellum, in contrast to the white scutellum of *Eriozona*. and the other two species differ in the venal pattern in the wing. *E. erratica* is a large hoverfly with a yellow pattern and is most easily confused with *Syrphus* and *Epistrophe*. The thoracic dorsum is blackish and shiny, in contrast to the more greenish dorsum of most species of the other two genera.

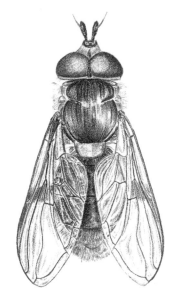

figure 309. *Eriozona syrphoides*, habitus of male (Verlinden).

Key

1.a. Body black, covered with long hairs, a bumblebee mimic (figure 309). Thorax has black hairs in the middle and yellow hairs on the front and hind margins. Scutellum white to yellow with yellow hairs. Abdomen white in front, black in the middle and has red hairs at the tip. Wing has dark patch in middle. 13-15 mm. Central Europe, in Asia east to Pacific **›** *Eriozona syrphoides* Fallén
1.b. Body black with yellow pattern, a wasp mimic without dense, long hairs (figure 310); scutellum yellow; wing hyaline, without dark patch. 10-15 mm. Northern and Central Europe, more montane in southern part of range, in Asia east to Pacific **›** *Eriozona erratica* Linnaeus (= *Megasyrphus erraticus* Zettersted, = *Megasyrphus annulipes* Zetterstedt)

ERISTALINUS

Introduction

Eristalinus fly in and above the vegetation, frequently basking in sunny spots. *E. aeneus* is regularly found hovering above bare soil or stones. Their behaviour is reminiscent of Muscid flies, but their eyes are covered with dark spots. The larvae have an elongated anal segment and live in a (semi)-aquatic environment, filtering bacteria and organic material.

Key

1.a. Tergites 2 and 3: with a dull spot; thoracic dorsum with 5 greyish stripes (figure 311, figure 312). Males: eyes separated on frons. Females: eyes hairy all over. 7-11 mm. Europe, North Africa, non-tropical part of Asia **›** *Eristalinus sepulchralis* Linnaeus
1.b. Tergites 2 and 3: completely shiny (figure 313), thoracic dorsum: in Northern and Central Europe with faint greyish stripes on front 1/2, but in Southern Europe with 5 strong greyish stripes. Males: eyes meet on frons (figure 314). Females: eyes bare on lower half. 10-12 mm. Cosmopolitan **›** *Eristalinus aeneus* Scopoli

figure 310. *Eriozona erraticus*, habitus of female (Verlinden).

figure 311. *Eristalinus sepulcralis*, habitus of male (Verlinden).

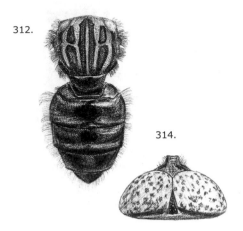

312.

314.

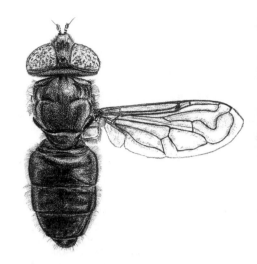

figure 312. *Eristalinus sepulcralis*, habitus of female.
figure 314. *Eristalinus aeneus*, head of male (Verlinden).

figure 313. *Eristalinus aeneus*, habitus of female (Verlinden).

ERISTALIS

Introduction

Eristalis can be found everywhere and many are common garden visitors. A limited number have quite specific habitat preferences and restricted distributions, for example *E. cryptarum*, which occurs in bogs. The larvae of *Eristalis* are detritivorus and bacteria-eating filter feeders, living in aquatic or semi-aquatic conditions, including liquified faeces. Larvae of *E. tenax* have incidentally been reported to occur in the intestinal tract, causing human myasis (maggot infestation).

Recognition

Most *Eristalis* can be identified by species specific characteristics, for example, *E. tenax* by its hair bands on the eyes and *E. similis* (= *E. pratorum*) by its long pale pterostigm. The colouration and extent of the yellow markings on the abdomen vary with temperature (Holloway, 1993). In particular, *E. arbustorum* and *E. horticola* extend the amount of yellow on the abdomen as the season progresses, more noticeably for males than for females. A number of species are easily confused (particularly in the field): firstly *E. arbusto-*

rum, E. abusiva and *E. nemorum*; secondly *E. horticola* and *E. jugorum* and thirdly *Eristalis picea, E. rupium, E. pseudorupium, E. alpina* and *E. nemorum*.
The key is based on Kanervo (1938), Nielsen (1995), Bartsch (1997), Zeegers and Van Veen (1992), Van Veen and Zeegers (1996), and Hippa et al. (2001). Nielsen (1995) clarifies the status of *E. fratercula, E. vallei* and *E. gomojunovae*. Hippa et al. (2001) clarify the status of *E. rupium, E. picea* and *E. pseudorupium*. The International Commission on Zoological Nomenclature restored the name *E. horticola* (formerly *E. lineata*) and *E. nemorum* (formerly *E. interrupta*).

Key

1.a. Arista bare or hairs shorter than twice the diameter of arista base (figure 315) ❭ 2
1.b. Arista feathery, hairs much longer than twice the diameter of arista (figure 316) ❭ 8

2.a. Eyes without bands of hair ❭ 3
2.b. Eyes with 2 distinct bands of hair (figure 317); arista bare. 14-16 mm. Cosmopolitan ❭ *Eristalis tenax* Linnaeus
Jizz: stout species with completely shiny abdomen. Femur 3 thick, hangs down while hovering.

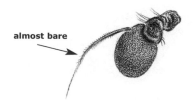

figure 315. *Eristalis tenax*, antenna (Verlinden).

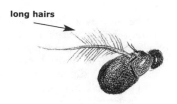

figure 316. *Eristalis arbustorum*, antenna (Verlinden).

figure 317. *Eristalis tenax*, head (Verlinden).

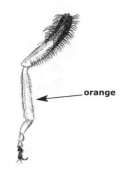

figure 318. *Eristalis cryptarum*, leg 1 (Verlinden).

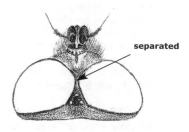

figure 319. *Eristalis abusiva*, head of male (Verlinden).

3.a. Tibia: at least tibia 3 dark at tip, often other tibia also dark at tip; antennae: 3rd segment black to brownish **>** 4

3.b. Tibia 1-3 entirely orange (figure 318); antennae: 3rd segment reddish. 10-12 mm. Northern and Central Europe, in Asia east to Mongolia **>** *Eristalis cryptarum* Fabricius
Jizz: whitish hind margins of tergites and orange legs, wings infuscate at front.

4.a. Tars 3 black; scutellum yellow (only whitish in *E. anthophorina*); tergites: hair colours not in white-black-red pattern, tergite 4 with white, brown or reddish hairs **>** 5

4.b Tars 3 orange; scutellum whitish; back of thorax and tergite 2 with white hairs, tergite 3 with black hairs, tergite 4 with red hairs. 12-15 mm. Northern Europe, south just into Central Europe (Baltic Sea) **>** *Eristalis oestracea* Linnaeus
Jizz: remarkably like *Eriozona syrphoides*.

5.a. Face with a broad, bare, black stripe; tibia blackish: black on top 1/2 or more; males: eyes meeting **>** 6
Jizz: scutellum pale, contrasts with abdomen.

5.b. Face without or (in older specimens) with a narrow black stripe; tibia pale: blackish on top 1/4 or less. Males: eyes separate (figure 319). 11-13 mm. **>** *Eristalis abusiva* Collin
Jizz: scutellum shiny, not contrasting much with the thoracic dorsum. Tibia 2 very pale, often with only a blackish patch at the very end, in contrast to *E. arbustorum*.

6.a. Face: below eyes pale-haired, in front view ground colour black or with yellowish parts; tergite 2: yellow spots (or place where they are normally present) dull or semi-shiny (compare with shiny parts of tergite 3, take care in strong light) **>** 7

6.b. Face: below eyes black-haired, ground colour black; tergite 2: yellow spots (or place where they are normally present) shiny. 12-15 mm. Northern parts of Palaearctic. ❯ *Eristalis fratercula* Zetterstedt (= *Eristalis vallei* auct.)

Jizz: dark band in wing. In males tergite 2 has yellowish basal corners and yellow hairs, tergite 3 dark-haired, tergites 4 and 5 white-haired, in females abdomen extensively yellowish-haired.

figure 320. *Eristalis similis*, habitus of male (Verlinden).

321.

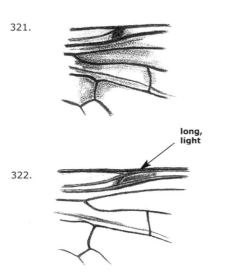

322.

long, light

figure 321. *Eristalis jugorum*, wing: stigma.
figure 322. *Eristalis similis*, wing: stigma (Verlinden).

7.a. Tergites 3 and 4: hairs predominantly pale, longer than width of metatars 3; face with pale yellow hairs. 11-13 mm. Northern parts of Palaearctic, up to the northern tree line. ❯ *Eristalis anthophorina* Fallén

Jizz: bumblebee mimic.

7.b. Tergites 3 and 4: large parts covered with black hairs, as long as the width of metatars 3; face with white, almost silver hairs. 8-11 mm. Northern Europe ❯ *Eristalis gomojunovae* Violovitsh

Jizz: bee mimic.

8.a. thorax and abdomen covered in short hairs, at most as long as width of metatars 3; squamulae whitish ❯ 9

8.b. thorax and abdomen covered in long hairs; squamulae greyish-black; postalar callus: hairs black. Males: abdomen completely brownish-haired. Females: abdomen with blackish hairs, last tergite with white hairs. 11-14 mm. Northern and Central Europe, increasingly montane further south, in Asia east to Siberia ❯ *Eristalis intricaria* Linnaeus

Jizz: the only densely haired *Eristalis* found south of Belgium, males' abdomen brownish-haired, females' black-haired with white tip.

9.a. Tars 1 and 2 at least with top segments black, often completely black ❯ 10

9.b. Tars 1 and 2 completely yellow; wing: with diffusely bordered darkened band in median part; pterostigma 4 times as long as wide. 12-16 mm. Europe ❯ *Eristalis pertinax* Scopoli

Jizz: males with triangular abdomen (figure 320), females with squarish abdomen.

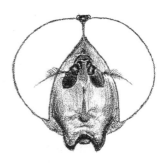

figure 323. *Eristalis arbustorum*, face of male (Verlinden).

324.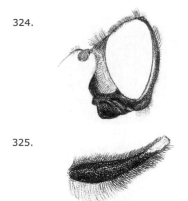

325.

326.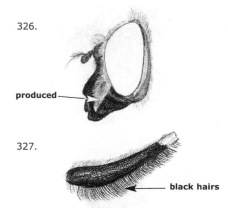

produced—

327.

black hairs

figure 324. *Eristalis picea*, head of male.
figure 325. *Eristalis picea*, femur 1
(Verlinden).

figure 326. *Eristalis jugorum*, head of male.
figure 327. *Eristalis jugorum*, femur 1
(Verlinden).

10.a. Pterostigm 1-3 times as long as wide, black (figure 321) **>** 11
10.b. Pterostigm 4-6 times as long as wide, yellow (females) or light grey (males) with dark border (figure 322). 13-15 mm. Europe, North Africa, east to Siberia **> *Eristalis similis*** Fallén (= *E. pratorum*)
Jizz: male with square, rather dark abdomen. Females appear much like *E. pertinax*.

11.a. Face: broad black stripe present from mouth edge to antennal bases **>** 12
11.b. Face: black stripe absent or very narrow, face entirely whitish (figure 323)[most old specimens have a narrow but distinct black stripe]; wing hyaline. 9-11 mm. Palearctic and Nearctics **> *Eristalis arbustorum*** Linnaeus
Jizz: small species, top 1/4 of tibia 2 black.

12.a. Mouth edge not elongated (figure 324). Male: without full row of black hairs at the back of femur 1, at most apical 1/2 black-haired (figure 325) **>** 13
12.b. Mouth edge projects downwards and forwards (figure 326); face: wide bare stripe; pterostigm black, square (figure 321); wing with weak black markings; femur 3 black at base (in contrast to *E. lineatus*). Male: femur 1 has a full row of dense, black hairs at the back (figure 327). Female: femur 1 with dense yellow hairs at the back. 11-13 mm. Elevated parts of Central Europe, in Asia east to Iran **> *Eristalis jugorum*** Egger

Jizz: yellow markings on abdomen distinctly orange-tinted, thorax with orange-reddish hairs, similar in appearance to *E. lineatus*.

13.a. Abdomen: tergite 2 partly shiny, at least on the yellow markings; tergites 3 and 4 with extensive band of black hairs in hind part **>** 14
13.b. Abdomen: tergite 2 completely dull, even on the yellow spots; tergites 3 and 4: entirely pale-haired or with a small patch of black hairs at hind margin; wing with dark markings, more developed in the female (figure 328); femur 3 pale at base (figure 329, figure 330, in contrast to *E. jugorum*). 11-14 mm. Europe, North Africa, in Asia to Pacific coast **> *Eristalis horticola*** Degeer (= *E. lineata* Harris)
Jizz: abdominal spots yellowish, the yellowish hind margins of tergites slightly more orange, but not as orange as *E. jugorum*, which it resembles.

figure 328. *Eristalis horticola*, habitus of male (Verlinden).

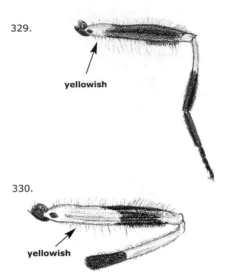

329.

yellowish

330.

yellowish

figure 329. *Eristalis horticola*, femur 3 of male.
figure 330. *Eristalis horticola*, femur 3 of
female (Verlinden).

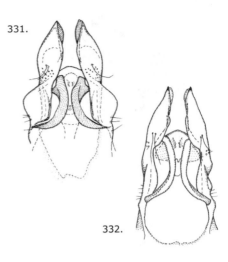

331.

332.

figure 331. *Eristalis hirta*, genitalia of male.
figure 332. *Eristalis alpina*, genitalia of male
(Hippa et al., 2001).

14.a. Thoracic dorsum dull with brownish to whitish dusting (if thorax shiny then metatars 3 pale); tergites 2-4: either with dull parts on tergites 2-4 or completely shiny (latter only females *E. rupium*); wing often darkened in middle. In some species: antennae brown and/or femur 3 pale at base **>** 15

14.b. Thoracic dorsum greenish black and shiny, at most in front part with very fine whitish pollinosity; metatars 3: black; antennae black; femur 3 black at base; wings hyaline. Males: tergite 2: pair of pale yellowish spots, tergites 3 and 4: black, completely shiny, at most tergite 3 with fine pollinosity on hind margin; genitalia as figure 331. Females: tergites 2-4 black, tergite 2 dull on middle, shiny otherwise, tergites 3 and 4 completely shiny. 10-13 mm. Circumboreal: extreme north of Palaearctic and Nearctic **>** *Eristalis hirta* Loew (= *Eristalis tundrarum* Frey)

Jizz: males with pale yellow spots, females with black abdomen and hyaline wings.

15.a. Femur 3: not thickened, at 2/3 of its length at most 1.5 times as thick as femora 1 and 2; tars 1 blackish; thoracic dorsum evenly and lightly dusted, no contrast between front and back **>** 16

15.b. Femur 3: thickened, at 2/3 of its length more than twice as thick as femora 1 and 2; tars 1 often with first 2-3 segments yellow, especially in the male; thoracic dorsum: front 1/2 heavily (female) or somewhat (male) dusted, in contrast to the black back 1/2; metatars 3 pale. Male: wing almost hyaline; male genitalia in figure 332. Female: square dark marking on wing; abdomen dark, at most tergite 2 with yellow markings, tergites dull only on hind part. 10-13 mm. Central Europe, elevated parts of Southern Europe, in Asia east to Pacific coast **>** *Eristalis alpina* (Panzer)

16.a. Pterostigm rectangular, longer than wide, extending to or beyond the point where vein sc merges into the wing border (figure 333); face: dusting thin, leaving a diffusely marked black stripe, facial knob often completely black and shiny; wing often with dark markings **>** 17

16.b. Pterostigm square (including surrounding veins even shorter than wide), its position basal to the point where vein sc merges into the wing border (figure 334); face: dusting dense, black stripe well-marked and relatively narrow; wings hyaline; male genitalia figure 335.**>** *Eristalis nemorum* Linnaeus (= *E. interrupta* Poda)

333.

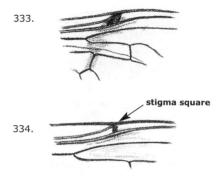

stigma square

334.

figure 333. *Eristalis picea*, wing: stigma of male.
figure 334. *Eristalis nemorum*, wing: stigma of male (Verlinden).

335.

tip of paramere

aedeagus

aedeagal lobe

336.

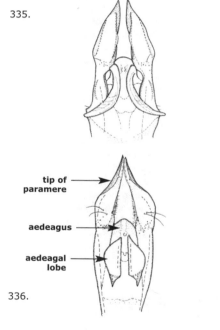

figure 335. *Eristalis nemorum*, genitalia of male.
figure 336. *Eristalis arbustorum,* genitalia of male (Hippa et al., 2001).

17.a. Male, to be keyed on genitalia ❯ 18
17.b.Females ❯ 20

18.a. Genitalia: tip of paramere narrowed; thoracic dorsum dull; normally only the very base of femur 3 pale ❯ 19

337.   tip of paramere

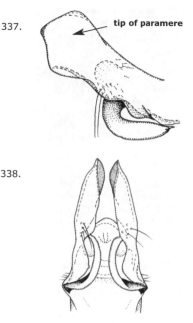

338.

figure 337. *Eristalis pseudorupium*, genitalia of male: hypandrium.
figure 338. *Eristalis pseudorupium*, genitalia of male (Hippa et al., 2001).

18.b. Genitalia: tip of paramere broad; aedeagal lobes strongly curved (fig. 338); thoracic dorsum shiny; normally base of femur 3 pale for 1/4. 10-13 mm. Northern Europe ❯ ***Eristalis pseudorupium*** Kanervo (= *Eristalis vitripennis* Strobl)

19.a. Genitalia: aedeagal lobes strongly curved (figure 339), tip of paramere pointed. Pterostigm about 1.5 times as long as wide (figure 340); subcosta often yellowish up to the pterostigm. 10-13 mm. Northern and Central Europe ❯ ***Eristalis picea*** Fallén
Jizz: dense, brown hairs on thorax.

19.b. Genitalia: aedeagal lobes weakly curved (figure 341); tip of paramere rounded; pterostigm about twice as long as wide (figure 342); subcosta often brownish-blackish on top 1/2. 10-13 mm. Northern Europe and elevated and mountainous parts of Central Europe, in Asia east to Siberia, Nearctic ❯ ***Eristalis rupium*** Fabricius
Jizz: dense, orange-reddish hairs on thorax.

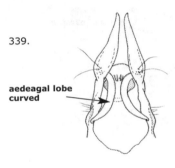

339.

**aedeagal lobe
curved**

340.

figure 339. *Eristalis picea*, genitalia of male
(Hippa et al., 2001).
figure 340. *Eristalis picea*, wing: stigma of
male (Verlinden).

341.

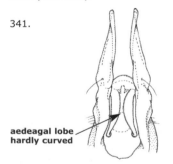

**aedeagal lobe
hardly curved**

342.

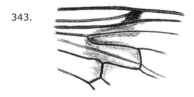

343.

figure 341. *Eristalis rupium*, genitalia of male
(Hippa et al., 2001).
figure 342. *Eristalis rupium*, wing: stigma of
male.
figure 343. *Eristalis picea*, wing: stigma of
female (Verlinden).

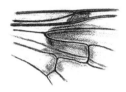

figure 344. *Eristalis rupium*, wing: stigma of
female (Verlinden).

20.a. Tergites 3 and 4 with a dull hind
border and often a dull front border;
pterostigm 1.5-3 times as long as wide
(figure 343); subcosta often yellowish up
to the pterostigm; wing with diffuse
blackish spot ❯ 21
20.b. Tergites 3 and 4 completely shiny,
with erect hairs; pterostigm 2-3 times as
long as wide (figure 344); subcosta often
brownish-blackish on top 1/2; wing
often with a well-marked blackish spot
❯ **Eristalis rupium** Fabricius

21.a. Tergite 3 with long, erect hairs;
pterostigm 2-3 times as long as wide;
metatars 3 black to yellowish; normally
the basal 1/4 of femur 3 pale, but femur
3 may be pale on basal 1/2 ❯ **Eristalis
picea** Kanervo
21.b. Tergite 3 posteriorly with short,
semi-erect to adpressed hairs; pterostigm
1.5-2 times as long as wide; in direct
comparison to *E. picea*, the eye hairs
sparser, shorter and paler; metatars 3 yellowish; normally basal 1/2 or more of
femur 3 pale ❯ **Eristalis pseudorupium**
Kanervo (= *Eristalis vitripennis* auct.)

EUMERUS

Introduction

Eumerus are small to medium-sized hoverflies. Most are thermophilous and can
be found in grasslands and at the borders
of woods and bushes, often near the larval
food plants. Most species fly quickly near
the ground, which makes them difficult
to spot. The larvae mine through bulbs
(e.g. *Allium*) and are considered pests in
commercial bulb plantations.

Recognition

Eumerus are either blackish flies with silvery spots on the tergites or reddish flies with a blackish thorax. Femur 3 is swollen and without a triangular tooth below the apex. There is no facial knob on the head and the mouth edge only projects slightly. The key is based on Van der Goot (1981), Seguy (1961) and study of specimens at the Zoological Museum of Amsterdam (ZMAN). *E. elaverensis, E. amoenus* and *E. pulchellus* are provisionally added. The first is added from the description of Seguy (1961), the latter two based on specimens in the ZMAN. All three are southern species that just reach the environs of Paris.

Key

1.a. Abdominal tergites 2 and 3 partly or completely red ❯ 2
1.b. Abdominal tergites 2 and 3 black, tergite 2 may contain a pair of yellow spots, tergites 3 and 4 may contain dust bands ❯ 14

2.a. Males, tip of abdomen with genital ball ❯ 3
2.b. Females, tip of abdomen pointed ❯ 8

3.a. Eyes bare ❯ 4
3.b. Eyes densely haired ❯ 5

4.a. Smaller: 5-7 mm. Eyes just separated on the frons (figure 345); ocelli in an isoscelese triangle, distance between the hind ocelli longer than between each hind to front ocellus (figure 345). 5-7 mm. Central and Southern Europe ❯ *Eumerus sabulonum* Fallén

4.b. Larger: 7-9 mm. Eyes meeting (figure 346); ocelli in a equilateral triangle: distance between ocelli equal (figure 346, variation exists!). 7-8 mm. Central and Southern Europe, North Africa, in Asia east to Mongolia ❯ *Eumerus tarsalis* Loew

5a. Abdomen not covered with silver-white, metallic hairs; black parts on thorax and head either blackish or with strong bronze sheen ❯ 6
5.b. Abdomen distinctively covered in silver-white, metallic hairs; black parts of thorax and head with strong blue sheen. 6-9 mm. Continental part of Central Europe and Southern Europe ❯ *Eumerus ovatus* Loew
Jizz: abdominal silver hairs cover tergite 3 and further.

6.a. Eyes separated, approaching each other at a single point (figure 347) ❯ 7
6.b. Eyes joined for some distance. 7-10 mm. Continental part of Central Europe, mountains of Southern Europe ❯ *Eumerus grandis* Meigen (= *Eumerus annulatus* Panzer)
Jizz: thorax blackish without metallic sheen.

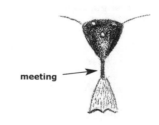

figure 346. *Eumerus tarsalis*, frons of male (Verlinden).

figure 345. *Eumerus sabulonum*, frons of male (Verlinden).

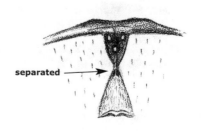

figure 347. *Eumerus tricolor*, frons of male (Verlinden).

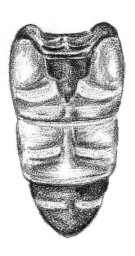

figure 348. *Eumerus tricolor*, abdomen of male (Verlinden).

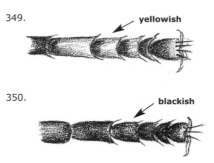

349. yellowish

350. blackish

figure 349. *Eumerus tarsalis*, tars 1 of female.
figure 350. *Eumerus tricolor*, tars 1 of female (Verlinden).

7.a. Thorax heavily punctuated, without shiny bronze sheen; mesonotic dorsum with short white to yellow hairs; abdomen figure 348. 7-10 mm. Continental part of Central Europe and Southern Europe, in Asia to Pacific coast **>** *Eumerus tricolor* Fabricius

7.b. Thorax hardly punctuate, with strong shiny bronze sheen; mesonotic dorsum with long brown to black hairs. 7-10 mm. Continental part of Central Europe **>** *Eumerus sinuatus* Loew

8.a. Tergites 3 and 4 with a black median stripe, only red at side margin **>** 9

8.a. Tergite 3 entirely red, tergite 4 entirely red or only red at margin **>** 11

9.a. Thorax and back of head either blackish or with bronze sheen **>** 10

9.b. Thorax and back of head with strong blue sheen. 6-9 mm. **>** *Eumerus ovatus* Loew

10.a. Thoracic dorsum blackish; tars 3 black; antenna: top of 3rd segment straight; femur 3 strongly thickened. 7-10 mm **>** *Eumerus grandis* Meigen (= *Eumerus annulatus* Panzer)

10.b. Thoracic dorsum with bronze sheen; tars 3 reddish-brown; antenna: top of 3rd segment rounded; femur 3 slightly thickened 7-10 mm. **>** *Eumerus sinuatus* Loew

11.a. Tars 1 and 2: first 3 segments ventrally white, the 2nd and 3rd with a basal black spot (figure 349) **>** 12

11.b. Tars 1 and 2 largely black to brown (figure 350) **>** 13

12.a. Smaller: 5-7 mm. Ocelli in a isosceles triangle, distance between the hind ocelli longer than between each hind to front ocellus (as figure 345) **>** *Eumerus sabulonum* Fallén

12.b. Larger: 7-9 mm. Ocelli about in a equilateral triangle: distance between ocelli about equal (as figure 346, variation exists!) **>** *Eumerus tarsalis* Loew

figure 351. *Eumerus sabulonum*, habitus of female (Verlinden).

13.a. Thoracic dorsum with short yellow hairs; tergite 4: entirely red **>** *Eumerus tricolor* Fabricius

13.b. Thoracic dorsum with long, erect white hairs; tergite 4: reddish only at margin **>** *Eumerus ovatus* Loew

14.a. Tergite 2 black, without yellow spots, but often marked with whitish dust spots. Male: tarsus 3 not broadened and flattened **>** 15

14.b. Tergite 2 with a pair of yellow spots, without dust spots (figure 352). Male: tarsus 3 broadened and flattened, with snow-white pilosity in front. 7 mm. Central and Southern Europe, in Asia to Pacific coast **>** *Eumerus flavitarsis* Zetterstedt

Jizz: black with bluish sheen.

15.a. Tergite 4 with reddish markings**>** 16

15.a. Tergite 4 black, at most with white dust stripes **>** 17

16.a. Males and females. Male: tergite 4 black with reddish hind margin; thoracic dorsum metallic, not blueish, shiny; tibia blackish, brown on middle, tars black above, reddish below; last tergites with silverish hair tufts at their margin. Female: tergite 4 narrowly red-dish; scutellum with spines at lateral margin reddish; antennal segment 3 reddish and large. 8-10 mm. Southwestern Europe, north to the environs of Paris **>** *Eumerus elaverensis* Seguy

16.b. Males only. Male: Tergite 4 black with reddish hind margin; thoracic dorsum black, blueish sheen; tibia black with yellowish base, tars black; hypopygium black-haired. 6-7 mm. Central and Southern Europe, North Africa **>** *Eumerus ruficornis* Meigen

17.a. Antennae orange to reddish, at least 3rd segment brown with base bright orange **>** 18

17.b. Antennae black, at most segment 3 with central area turning brown **>** 27

18.a. Males, eyes meeting **>** 19

18.b. Females, eyes separated **>** 23

19.a. Male: sternite 3 flat, without projection and without tufts of long white hairs [cerci large and globular for *E. ornatus*] **>** 20

19.b. Male: sternite 3 with a low projection on front 1/2, implanted with tufts of long white hairs, in one species acompanied by a thorn-like high projection in hind part [cerci never globular] **>** 22

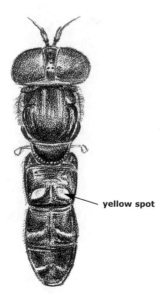

yellow spot

figure 352. *Eumerus flavitarsis*, habitus of female (Verlinden).

figure 353. *Eumerus ornatus*, habitus of male (Verlinden).

20.a. Tergites 3 and 4: hairs along their side margin equally long to the hairs on the tergite disk; hypopygium black-haired ❭ 21

20.b. Tergites 3 and 4 with long, downward pointing white hairs, that are more than twice as long as the hairs on the tergite; hypopygium white-haired; ocelli located on the middle of the vertex, distance between front ocellus and both hind ocelli approximately equal to or somewhat shorter than the distance between hind ocelli and hind margin of the eyes; cerci not large, globular and yellowish, not visible from below. 7-10 mm. Southern Europe, extending north to the Paris basin. *Eumerus pulchellus* Loew

21a Ocelli located on the front of the vertex, distance between front ocellus and both hind ocelli about 1/2 the distance between the hind ocelli and the hind margin of the eyes (figure 353); cerci large and globular, yellowish, visible from below; 7-10 mm. Central and Southern Europe, North Africa ❭ *Eumerus ornatus* Meigen

Jizz: elongate *Eumerus* with large eyes, frons very small.

Note: If only the 3rd antennal segment is brownish/reddish in the middle (many females) or the male has a white-haired hypopygium: *E. strigatus* [normally antennae black, but they may turn brownish].

21b Ocelli located on the middle of the vertex, distance between front ocellus and both hind ocelli equal to or somewhat longer than the distance between hind ocelli and hind margin of the eyes; cerci not large, globular and yellowish, not visible from below. 7-10 mm. Southern Europe, extending north to the Paris basin. ❭ *Eumerus amoenus* Loew

Jizz: black abdomen with sharply-defined white dust spots.

22.a. Sternite 3: front 1/2 with low projection, implanted with tufts of long white hairs, hind 1/2 with a triangular, thorn-shaped projection, embedded with black hairs. 7-8 mm. Continental part of Central Europe and Southern Europe, North Africa ❭ *Eumerus clavatus* Becker

22.b. Sternite 3: only a low projection on front 1/2, implanted with tufts of long white hairs. 7-8 mm. Continental part of Central Europe ❭ *Eumerus uncipes* Rondani

23.a. Coxa 1 entirely yellow ❭ 24

23.b. Coxa 1 largely blackish: black, black with yellow apex or yellow with a large blackish patch ❭ 25

24.a. Ocelli on the front part of the vertex, the distance between the front and hind ocelli shorter than the distance from the hind ocelli to the line that connects the hind corners of the eyes. 8 mm. Central and Southern Europe, North Africa ❭ *Eumerus ornatus* Meigen

24.b. Ocelli located on the hind part of the vertex, the distance between the front and hind ocelli is larger than the distance from the hind ocelli to the line that connects the hind corners of the eyes. 6-7 mm. Central and Southern Europe, North Africa ❭ *Eumerus ruficornis* Meigen

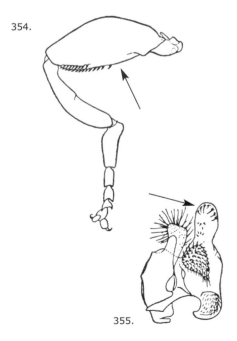

354.

355.

figure 354. *Eumerus funeralis*, leg 3.
figure 355. *Eumerus funeralis*, genitalia of male (after Van der Goot, 1981).

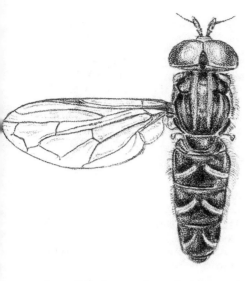

figure 356. *Eumerus funeralis*, habitus of male (Verlinden).

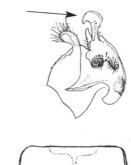

357.

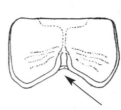

358.

figure 357. *Eumerus strigatus*, genitalia of male.
figure 358. *Eumerus strigatus*, sternite 4 of male (after Van der Goot, 1981).

25.a. Tars 1 blackish above or tarsal segments yellowish with blackish patches, tars 2 and 3 largely blackish above **>** 26
25.b. Tars 1 entirely yellow above, tars 2 and 3 largely yellow above **> *Eumerus clavatus*** Becker and ***Eumerus uncipes*** Rondani
Note: females cannot be distinguished.

26.a. Frons: surface flat, with short, whitish adpressed hairs on median part and shiny margins, finely punctuate with diameter of dots much shorter than distance between dots, vertex shiny **> *Eumerus amoenus*** Loew
26.b. Frons: surface wrinkled, entirely dusted, vertex lightly to densely dusted. 7-10 mm. Southern Europe, reaching north to the Paris basin. ***Eumerus pulchellus*** Loew

27.a. Femur 3: projection absent, ventrally covered with long dense hairs over the full length **>** 28
27.b. Femur 3: at the base ventrally a small projection (do not mistake trochanter for this projection!), apical to this projection a bare, shiny area (figure 354). Male: genitalia figure 355. 5-6 mm. Cosmopolitan in warmer areas worldwide **> *Eumerus funeralis*** Meigen
(= *Eumerus tuberculatus* Rondani)
Jizz: black with slight greenish shine.

359.

360.

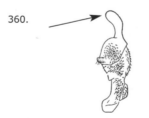

361.

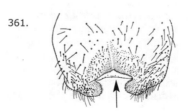

figure 359. *Eumerus sogdianus*, head of male, top view.
figure 360. *Eumerus sogdianus*, genitalia of male.
figure 361. *Eumerus sogdianus*, sternite 4 of male (after Van der Goot, 1981).

28.a. Males: ocelli in a isoscelese triangle; genitalia: surstyli hooked (figure 357); sternite 4 without lobular projections at the hind margin (figure 358). Females: indistinguishable from the next. 5-6 mm. Europe, in Asia east to Japan **>** *Eumerus strigatus* Fallén

28.b. Male: ocelli in a equilateral triangle (figure 359); genitalia: surstyli straight (figure 360); sternite 4 with lobular projections at the hind margin (figure 361), longer than wide. Females: indistinguishable from the previous. 7-8 mm. Central and Southern Europe, in Asia east to Mongolia **>** *Eumerus sogdianus* Stackelberg

EUPEODES

Introduction

Eupeodes (= Metasyrphus) species are medium-sized hoverflies with a black abdomen with yellow markings. Some of the species are particularly common, for instance *E. corollae*, while others are localised, for instance *E. nielseni*. *E. latifasciatus* tends to migrate and can be found in many places. *Eupeodes* larvae prey on aphids.

Recognition

The species of *Eupeodes* are hard to separate. Colour patterns may vary depending on the temperatures during pupation. Specimens will be darker at lower and lighter at higher temperatures during the pupal stage. Nevertheless, the present key uses some of the classic colour characteristics as no alternative is available.

The microtrichia coverage of the wing is used in several places. It is more stable than colour characteristics. Two sources of error exist: (1) small bare areas may be formed during preparation of specimens, and (2) some variation exists. Females always have a lower microtrichia coverage than males and it is necessary to match the sex when comparing specimens.

The key is based on Dusek and Laska (1976), Mazanek et al. (1999b), Mazanek et al. (1999c), Nielsen (2003), Mazanek et al. (2004) and the key of Zeegers and Van Steenis (determinatiemap). The key is provisional with regard to *E. curtus, E. duseki* and *E. goeldlini. E. duseki* and *E. goeldlini* were included from the descriptions of Mazanek et al. (1999c), *E. curtus* is included from Vockeroth (1992) and Mazanek et al. (1999b).

Key

1.a. Alula partly bare, partly covered with microtrichia (figure 362, figure 363)**>** 2
1.b. Alula entirely covered with microtrichia **>** 8

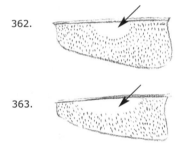

figure 362. *Eupeodes nielseni*, wing: alula.
figure 363. *Eupeodes luniger*, wing: alula (Verlinden).

2.a. Vein R4+5 almost straight; metasternum hairy; tergites 3 and 4: spots may or may not reach side margin **>** 3
2.b. Vein R4+5 strongly bent (figure 364); metasternum bare; tergites 3 and 4: spots do not reach side margin. 8-12 mm. Europe, in Asia to Pacific coast, Nearctic **>** *Eupeodes lapponicus* Zetterstedt
Jizz: as dark and large as *Dasysyrphus pinastri*, but eyes bare.

figure 364. *Eupeodes lapponicus*, wing (Barendregt, 2001).

3.a. Smaller species: 7-11 mm; wing: more than 50% (often 70% or more) of the wing membrane covered with microtrichia. Male: eye without enlarged facets; face above antennae not swollen; angle between eyes often 90° or less, at most 110° (*E. nielseni* and *E. biciki*). Female: wing also partially or entirely covered with microtrichia in basal 1/3 of wing **>** 4

3.b. Large species: 11-12 mm; wing: less than 40% of the wing membrane covered with microtrichia. Male: eye: upper 2/3 with distinctly enlarged facets; face above antennae swollen, the hind 1/2 densely black-haired; angle of approximation between eyes very wide (120°). Female: wing without microtrichia in basal 1/3 of wing. Northern Europe and northern parts of Central Europe, in Asia east to Pacific **>** *Eupeodes lundbecki* Soot-Ryen

Jizz: similar to *Scaeva*, but eyes bare.

4.a. Sternites 3 and 4 with large, rectangular black spots which may connect (figure 365); femur 1: long black hairs at the base. Male: eyes: angle of approximation on frons 105-110° (figure 366); postocular orbit near vertex 0.3-0.5 times its maximal width. Female: frons above lunulae with 2 dark spots **>** 5

4.b. Sternites 3 and 4: spots oval or rounded, often small and never connected; femur 1: long yellow hairs at the back of the base. Male: eyes: angle of approximation less than 100°; postocular orbit near vertex more than 0.5 times its maximal width. Female: frons directly above lunulae yellow **>** 6

5.a. Scutellum: majority of hairs black; tergite 5: side margin largely to completely black. Male: postocular orbit near vertex very narrow: less than 0.3 of its maximal width, with a row of black hairs near vertex. Female: frons above lunulae with 2 dark spots; femur 3 black on basal 1/2. 8-10 mm. Northern and Central Europe **>** *Eupeodes nielseni* Dusek & Laska

365.

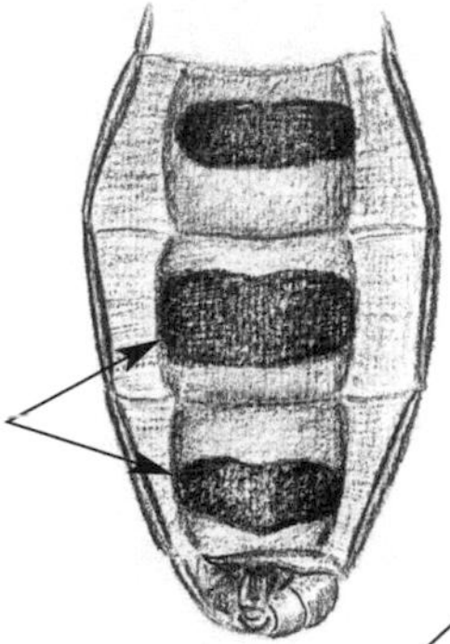

366.

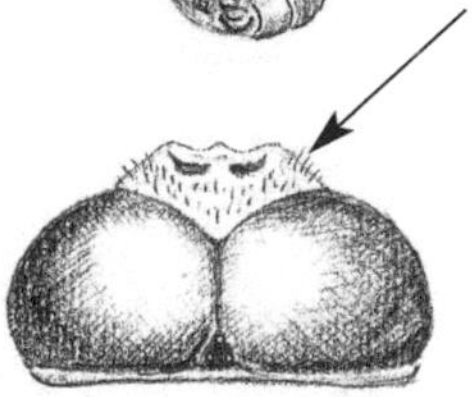

367.

figure 365. *Eupeodes nielseni*, sternites.
figure 366. *Eupeodes nielseni*, head of male, top view.
figure 367. *Eupeodes nielseni*, abdomen of male (Verlinden).

figure 368. *Eupeodes biciki*, abdomen of male (Nielsen, 2003).

5.b. Scutellum: hairs yellow; tergite 5: side margins broadly yellow. Male: postocular orbit near vertex broader: a little narrower than 0.5 of its maximal width, completely yellow-haired. Female: not described, but most probably characterised by the yellow hairs on the scutellum, the long black hair behind femora 1 and 2 and the broad face (Nielsen, 2003). 8-9 mm. Subarctic areas of Scandinavia. **> Eupeodes biciki** Nielsen

6.a. Tergites 3 and 4: spots do not reach side margin, or at most at their extreme front end (figure 377); scutellum: predominantly black-haired. Male: genitalia small, reaching hind margin of sternite 5 (figure 372). Female: frons either without dust spots or with a Y of black extending into the yellow **>** 7

6.b. Tergites 3 and 4: spots reach the side margin (figure 369, figure 370); scutellum: typically yellow-haired, but some black hairs may be present. Male: genitalia large, reaching hind margin of sternite 4 (figure 371). Female: frons with dust spots and border between black and yellow straight (figure 373). 6-10 mm. Europe, Africa, Asia **> Eupeodes corollae** Fabricius

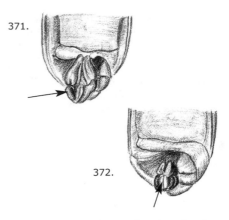

371.

372.

figure 371. *Eupeodes corollae*, last sternites and genitalia of male.
figure 372. *Eupeodes latifasciatus*, last sternites and genitalia of male (Verlinden).

figure 373. *Eupeodes corollae*, head of female (Verlinden).

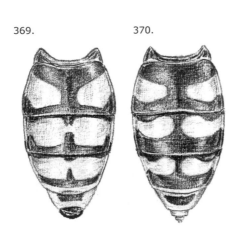

369.   370.

figure 369. *Eupeodes corollae*, abdomen of male.
figure 370. *Eupeodes corollae*, abdomen of female (Verlinden).

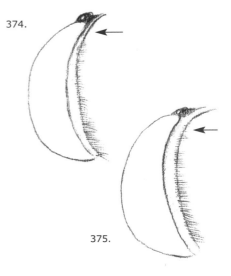

374.

375.

figure 374. *Eupeodes luniger*, head of male.
figure 375. *Eupeodes latifasciatus*, head of male (Verlinden).

7.a. Male: postocular orbit small near vertex, less than 0.5 of its maximal width (figure 374). Female: frons with dust spots, the visible black part extends as a black Y into the yellow part (figure 376). Both sexes (but variable): tergite 5: side margin yellow. 8-12 mm. Europe, North Africa, in Asia east to Japan **>** *Eupeodes luniger* Meigen

7.b. Male: postocular orbit broad near vertex, more than 0.5 of its maximal width (as figure 375). Female: frons without dust spots, border between black and yellow straight. Both sexes (but variable): tergite 5: side margin mainly black (figure 377 figure 378). 7-11 mm. Boreomontane species **>** *Eupeodes tirolensis* Dusek and Laska

8.a. Males and females. Female: frontal dust spots present, lateral margin of tergite 5 yellow or black; scutellum yellow- or black-haired **>** 9

8.b. Females. Female: frontal dust spots absent, frons entirely shiny; lateral margins of tergite 5 always yellow; tergites 3 and 4 normally with spots connected as broad bands reaching lateral margins, in dark specimens (e.g. occurring in spring) the spots separate; scutellum predominantly with yellow hairs. 7-9 mm. Europe, North Africa, in Asia east to Pacific, Nearctic **>** *Eupeodes latifasciatus* Macquart

Jizz: shiny frons of female distinctive in the field.

9.a. Sternites 3 and 4: black spots of equal size, clearly marked, rectangular or nearly so, sometimes with lateral margin pointing forward (figure 379); front margin of spots or bands not parallel to tergite front margin. Males: postocular margin less than 0.5 times its maximal width near vertex. Female: vertex narrow, as broad as the distance between the lateral outlines of the antennal sockets (figure 381) **>** 10

376.

377.

378.

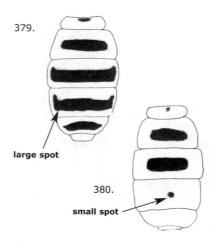

379.

large spot

380.

small spot

figure 376. *Eupeodes luniger*, head of female, top view.

figure 377. *Eupeodes luniger*, abdomen of male.

figure 378. *Eupeodes luniger*, abdomen of female (Verlinden).

figure 379. *Eupeodes nitens*, sternites of female.

figure 380. *Eupeodes punctifer*, sternites of female (Mazanekl et al., 2004).

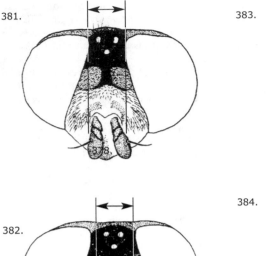

381.

382.

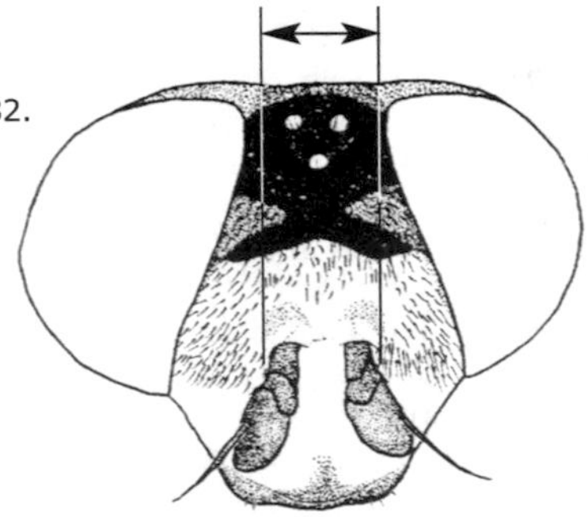

383.

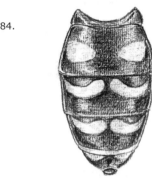

384.

figure 381. *Eupeodes curtus*, head of female with narrow vertex.
figure 382. *Eupeodes abiskoensis*, head of female with broad vertex (Mazanek et al., 2004).

figure 383. *Eupeodes nitens*, abdomen of male.
figure 384. *Eupeodes nitens*, abdomen of male (Verlinden).

9.b. Sternites 3 and 4: blackish spot on sternite 4 smaller than that on sternite 3 or even lacking, rounded to oval (figure 380), if spots on sternite 4 approach the size of those on sternite 3 (in males of *E. latifasciatus*), then postocular orbit broad, near vertex more than 0.5 times its maximal width and front margin of spots/bands on tergites 3 and 4 straight, parallel to front margin of tergite. Female: vertex broad, distinctly broader than the distance between the lateral outlines of the antennal sockets (figure 382) ❭ 14

10.a. Tergite 5: lateral margin mainly yellow. Male: postocular orbit near vertex more than 0.3 times its maximal width. Female: femur 3 yellow or with black base. Both sexes (but variable): sternites 3 and 4: spots oval or rectangular with rounded corners ❭ 11

10.b. Tergite 5: lateral margin entirely black, tergites 3 and 4 normally with bands, which may separate (figure 383, figure 384); 2nd basal cell br covered with microtrichia for about 50%. Male: postocular orbit near vertex very small, less than 0.3 times its maximal width. Female: femur 3 yellow. Both sexes (but variable): long hairs on back of base of femur 1 all black; sternites 3 and 4: spots strongly rectangular with angular corners, forming bands. 8-11 mm. Northern and Central Europe, in Asia east to Pacific ❭ *Eupeodes nitens* Zetterstedt

11.a. Wing: basal cells bm and br covered for 50-100% with microtrichia; tergites 3 and 4: spots or band may or may not reach the side margin. Male (*E. duseki, E. goeldlini*): genitalia: paramere with basal tooth long and not significantly overlapping the periphery of the paramere in lateral view ❭ 12

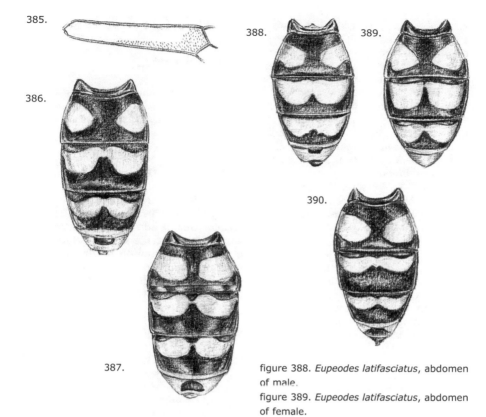

385.

386.

387.

388.

389.

390.

figure 385. *Eupeodes bucculatus*, wing: cell bm.
figure 386. *Eupeodes bucculatus*, abdomen
of female.
figure 387. *Eupeodes bucculatus*, abdomen
of male (Verlinden).

figure 388. *Eupeodes latifasciatus*, abdomen
of male.
figure 389. *Eupeodes latifasciatus*, abdomen
of female.
figure 390. *Eupeodes latifasciatus*, abdomen
of female (Verlinden).

11.b. Wing: basal cells bm and br covered for 50% or less with microtrichia (figure 385), (but alula fully covered with microtrichia); tergites 3 and 4: spots or bands reach the side margin in the front corner; sternites 3 and 4: spots rectangular or elongate oval. Male: genitalia: paramere with basal tooth shorter, about as long as wide. 9-12 mm. Atlantic part of Europe 〉 *Eupeodes bucculatus* Rondani (= *Eupeodes latilunulatus* Collin)

12.a. Wing: 2nd basal cell br covered for 50-90%, in rare cases almost 100%, with microtrichia. Male: genitalia: base of aedeagus with a strong tooth and pointed tip; hypandrium, ventral view: cavity deep, about 1/3-1/2 hypandrium height 〉 13

12.b. Wing: 2nd basal cell br completely covered with mictrotrichia. Male: genitalia: base of aedeagus with 2 teeth, tip also extending into a tooth; hypandrium, ventral view: cavity shallow, about 1/5 - 1/4 hypandrium height. 8-12 mm. Northern Europe, Nearctic 〉 *Eupeodes curtus* Hine

13.a. Tergites 3 and 4: marked with spots that do not reach the side margin; wing: 2nd basal cell covered for 90-100% (male) or 60-90% (female) with microtrichia. 8-12 mm. Northern Europe 〉 *Eupeodes duseki* Mazanek, Laska & Bicik

13.b. Tergites 3 and 4: marked with bands that reach the lateral margin; wing: 2nd basal cell covered for 50-85% (male) with microtrichia. Female unknown. 8-12 mm. Central Europe 〉 *Eupeodes goeldlini* Mazanek, Laska & Bicik

14.a. Males and females. Tergites 3 and 4: yellow spots separated, upper margin concave and spots well-separated from the front margin of the tergite; scutellum predominantly with black hairs on disk. Female: frons with dust spots ❯ 15

14.b. Males. Tergites 3 and 4: yellow spots connected, rarely separated and then front margin almost straight, parallel with tergite front margin and very close to it (figure 388, figure 389, figure 390); scutellum predominantly with yellow hairs on disk. 7–9 mm. Europe, North Africa, in Asia east to Pacific, Nearctic ❯ *Eupeodes latifasciatus* Macquart
Jizz: males with large 'ski-glasses' on tergites.

15.a. Wing: 2nd basal cell covered with microtrichia for more than 50%. Male: postocular orbit behind head not narrowed and very broad. Female: hairs on scutellum rather short and brush-like, straight in apical part; thoracic dorsum with hairs yellowish to golden-yellow ❯ 16

15.b. Wing: 2nd basal cell covered with microtrichia for less than 50%; face: a black stripe from the facial knob downwards. Male: postocular orbit distinctly narrowed near vertex. Female: hairs on scutellum rather long and fine, at least some hairs on scutellar margin tortuous in apical part; thoracic dorsum with hairs pale yellow. 7–11 mm. Northern Europe ❯ *Eupeodes punctifer* Frey (= *Eupeodes chillcotti* Fluke)
Jizz: rather dark species, with relatively narrow yellow spots.

16.a. Male: postocular orbit very broad; tip of the mid part of aedeagus elongate, twice as high as broad. Female: black spot on tergite 5 small or absent. 7–10 mm. Northern Scandinavia ❯ *Eupeodes abiskoensis* Dusek & Laska

16.b. Only males known. Postocular orbit smaller; tip of the mid part of aedeagus less elongate. 8 mm. Northern Scandinavia ❯ *Eupeodes borealis* Dusek & Laska

FERDINANDEA

Introduction

Ferdinandea are found along forest edges, where adults perch on the bark of trees and visit flowers. *Ferdinandea* larvae live in the sap streams of deciduous trees.

Recognition

Ferdinandea are reminiscent of muscid flies in the way that they sit on bark. The strong black bristles on the thorax and scutellum add to the similarity. However, the body colouration easily identifies them as *Ferdinandea*: the thorax is greyish, the abdomen is greenish and there are dark spots on the wings.

Key

1.a. Arista black, head as figure 392; tergites: dull bands connect to hind margin of tergites; thoracic dorsum black with grey dust, with longitudinal black stripes (figure 391). 10–13 mm. Europe, North Africa, in Asia east to Japan ❯ *Ferdinandea cuprea* Scopoli

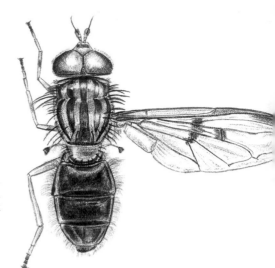

figure 391. *Ferdinandea cuprea*, habitus of male (Verlinden).

1.b. Arista yellow to red, head as figure 393; tergites: dull bands clearly separated from hind margin of tergites; thoracic dorsum black with grey dust, with longitudinal black stripes. 10-12 mm. Central Europe, in Asia east to Northern China **> Ferdinandea ruficornis** Fabricius

392.

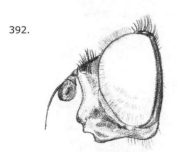

393.

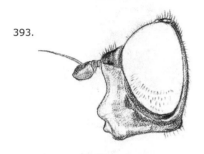

figure 392. *Ferdinandea cuprea*, head of male.
figure 393. *Ferdinandea ruficornis*, head of male (Verlinden).

HAMMER-SCHMIDTIA

Introduction

Hammerschmidtia inhabit coniferous (*Pinus*) or deciduous (*Betula, Quercus*) forest with old growth *Populus tremula*. The adults may be found sitting on the trunks of *Betula* and old *P. tremula*, or on logs and stumps in the vicinity of *P. tremula*. They visit the flowers of white umbellifers, *Ranunculus*, and male *Salix* (Speight, 2003).

Recognition

Hammerschmidtia are large orange-brown flies, which resemble Scatophagidae rather than Syrphidae. They might be confused with *Brachyopa*, but *Hammerschmidtia ferruginea* is significantly larger, with an elongate abdomen.

Key

1. Single European species. Large, elongate, orange-brown hoverfly; arista feathered. 10-12 mm **>** *Hammerschmidtia ferruginea* Fallén

HELOPHILUS

Introduction

Helophilus are large hoverflies that prefer moist and wet habitats. They visit flowers with shallow nectaries and bask on leaves in the sun. Most of them are good flyers and they may disperse across large distances. *H. affinis* migrates a long way south and may occur as far south as The Netherlands.
Their larvae are semi-aquatic filter feeders on detritus and bacteria. They are associated with accumulations of decaying organic material in ponds and mud. Their last segment is elongated into a tail, like that of *Eristalis* (Rotheray, 1993).

Recognition

Large, yellow and black hoverflies, with longitudinal whitish stripes on the black thoracic dorsum. Spots on tergite 4 often grey. The key is based on Van der Goot (1981), Verlinden (1991) and Nielsen (1997).

Key

1.a. Tergite 3 with yellow spots; tergite 2: yellow spots large, they cover most of the side margin **〉 2**

1.b. Tergite 3: without yellow pattern, but black with small grey band, which may be interrupted in the middle, tergite 2 with narrow yellow spots, which cover less than half the side margin (figure 394). 10-12 mm. Northern Europe **〉** *Helophilus bottnicus* Wahlberg

2.a. Thoracic dorsum without a band of black hairs between wings, predominantly with pale hairs; scutellum pale haired; tergites: hind margins black or yellow **〉 3**

2.b. Thoracic dorsum: a wide band of black hairs between the wings and pale hairs on other parts; scutellum black haired; tergites 2 and 3: yellow spots end medially in a grey spot; tergites 3 and 4: hind margin black; face: bare stripe blackish. 11-13 mm. Northern Europe, in Asia east to Pacific, Nearctic **〉** *Helophilus groenlandicus* Fabricius
Jizz: yellow markings on tergite 2 sharply pointed at their inner corner, with a grey spot.

3.a. Thoracic dorsum: longitudinal lines of pale dust reach the hind margin over their full width; tergite 4 always with yellow or white markings **〉 4**.

3.b. Thoracic dorsum: longitudinal lines of pale dust fade towards the scutellum and do not reach the hind margin of the thoracic dorsum; tergite 2 with large yellow spots, tergite 3 with much smaller yellow spots and tergite 4 completely black or with spots of white dust (in the latter case formerly recognized as *H. borealis*); tergites 3 and 4: hind margin

figure 394. *Helophilus bottnicus*, abdomen of female (after Van der Goot, 1981).

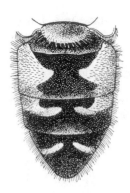

395.

396.

figure 395. *Helophilus lapponicus*, abdomen of male.
figure 396. *Helophilus lapponicus*, abdomen of female (after Van der Goot, 1981).

black (figure 395, figure 396); face: bare stripe blackish. 11-13 mm. Northern Europe, Nearctic **〉** *Helophilus lapponicus* Wahlberg (= *Helophilus borealis* Staeger)
Jizz: quite like *Eristalis* at first sight, because of short and broad abdomen and placement of spots.

4.a. Face: longitudinal bare stripe black (figure 397). Females: tergite 3: yellow spots separated from front margin medianly **〉 5**

4.b. Face: longitudinal bare stripe yellow (figure 398); tergites 2 and 3 with lemon yellow spots, tergite 4 with white to grey spots (figure 399, figure 400). Females: tergite 3: yellow spots touch front margin. 15-16 mm. Europe, in Asia east to Pacific **〉** *Helophilus trivittatus* Fabricius
Jizz: large species, spots on tergites 2 and 3 very pale, tergite 4 always with grey dust spots.

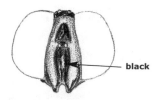

black

figure 397. *Helophilus pendulus*, face (Verlinden).

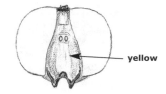

yellow

figure 398. *Helophilus trivittatus*, face (Verlinden).

5.a. Tergites 3 and 4: hind margin with a yellow band; tibia 1: top 1/3 completely black, not fading to yellow on inner side; tars 1 completely black above and strongly darkened to dark brown below **〉 6**

5.b. Tergites 3 and 4: hind margin black, at most turning vaguely yellow in median part; tibia 1: outer part of top 1/4 black, fading to yellow at the inner part. Males: tars 1 pale; tibia 3: base yellow for 1/3-1/2. Females: tars 1: pale, but often darkened above and yellow below; tibia 3: base yellow for 1/3. 14-16 mm. Northern Europe, northern part of Central Europe, east to Siberia **〉** *Helophilus affinis* Wahlberg

Jizz: tibiae and tarsae extensively pale. Males: spots on tergite 2 lemon yellow, those on tergite 3 more yolk yellow. Females similar to *H. hybridus*, but body broader and legs paler.

6.a. Femur 3: pale on the top 1/3-1/2; tibia 3: yellow on basal 1/2 or more. Males: yellow spots reach the hind margin only at the side corner, leaving a large black part between the spot and the hind margin (figure 401). Female: abdomen figure 402. Smaller species: 11-13 mm **〉** *Helophilus pendulus* Linnaeus

Jizz: smaller species, spots on abdomen yolk yellow, grey dust spots on tergite 4 often replaced by normal pale yellow spots.

6.b. Femur 3: at most with top 1/4 yellow, often less; tibia 3: pale on basal 1/4 or less. Males: tergite 2: yellow spots reach the hind margin over their full width (figure 403). Female: abdomen figure 404. Larger species: 14-16 mm **〉** *Helophilus hybridus* Loew

Jizz: males unique due to large yellow spots on tergites 2 and 3. Grey dust spots on tergite 4 well-developed. Females: stout *Helophilus*, with straight yellow spots on tergites 3 and 4, like *affinis*, but legs darker.

399.

400.

figure 399. *Helophilus trivittatus*, abdomen of male.
figure 400. *Helophilus trivittatus*, abdomen of female (after Van der Goot, 1981).

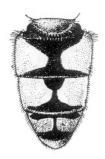

401.

402.

403.

404.

figure 401. *Helophilus pendulus*, abdomen of male.
figure 402. *Helophilus pendulus*, abdomen of female.
figure 403. *Helophilus hybridus*, abdomen of male.
figure 404. *Helophilus hybridus*, abdomen of female (after Van der Goot, 1981).

HERINGIA

Introduction

Heringia are often encountered in dense vegetation in forest edges. They have a swift erratic flight pattern, similar to small black bees, e.g. *Halictus*. *Neocnemodon* is merged into *Heringia* in this key. The larvae of *H. heringi* are predators of gall-forming or leaf-curling aphids and psyllids on various trees, e.g. *Populus* and *Ulmus*, and fruit trees e.g. *Malus*, *Prunus*, and *Pyrus*. They live within the aphid galls, one larva per gall being normal (see (Speight, 2003) and the reference therein). The larvae of the subgenus *Neocnemodon* prey upon coccids, aphids and adelgid plant bugs on coniferous trees, e.g. *Picea* and *Pinus*.

Recognition

Heringia are small, black hoverflies close to *Pipiza, Pipizella* and *Trichopsomyia*. In *Heringia*, the abdomen is elongated and the tergites are completely black, in contrast to most *Pipiza* and *Trichopsomyia*. *Heringia* is close to *Pipizella*, which has also an elongated third antennal segment. In *Heringia*, vein tp ends in a sharp angle on R4+5 in the wing, in *Pipizella* vein tp ends perpendicular to R4+5. In addition, *Heringia* are blackish at first sight, while *Pipizella* look brownish.

Key

1.a. Antennae: 3rd segment round, as long as broad (figure 405), in the female somewhat swollen. Male: coxa 2 and trochanter 3 with elongated protuberances **>** 2

1.b. Antennae: 3rd segment elongate, 2-3 times as long as broad (figure 406, figure 407). Male: coxa 2 and trochanter 3 without protuberances. 5-7 mm. Central and Southern Europe, in Asia to Mongolia **>** ***Heringia (Heringia) heringi*** Zetterstedt

Note: *Heringia senilis* is treated as a synonym.

405.

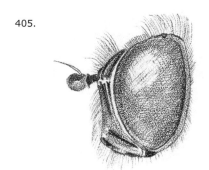

408.

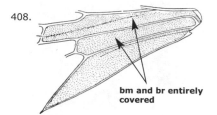

bm and br entirely covered

409.

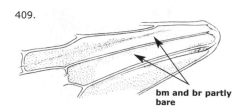

bm and br partly bare

figure 408. *Heringia pubescens*, wing: basal cells.
figure 409. *Heringia vitripennis*, wing: basal cells (after Van der Goot, 1981).

406.

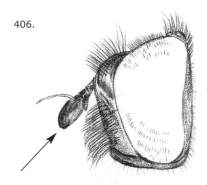

407.

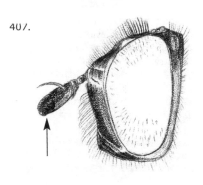

figure 405. *Heringia brevidens*, head of male.
figure 406. *Heringia heringi*, head of male.
figure 407. *Heringia heringi*, head of female (Verlinden).

2.a. Wing: basal cells bm and br and anal cell entirely covered by microtrichia (figure 408); smaller species: 5-7 mm **>** 3
2.b. Wing: basal cells bm and br and anal cell bare at least on first 1/2 (figure 409, microtrichia often present along the venia spuria); mostly larger species: 6-8 mm, one small species 4.5-5 mm. Only males can be identified. **>** 4

3.a. Males: sternite 4 with a small prominence, sternite 3 flat; metatars 1 cylindrical, without a depression at the back, tibia 2 without keel; thoracic dorsum mainly white-haired. Females: eye hairs sparse, inconspicuous and pale. 5 mm. Early spring species. Central Europe, in Asia to Pacific coast **>** *Heringia (Neocnemodon) verrucula* Collin
Jizz: small, black species, legs thin. In early spring on willow catkins.
3.b. Males: sternite 4 flat, without prominence, sternite 3 flat; metatars 1 with a depression at the back (figure 410), tibia 2 with keel (figure 411); thoracic dorsum black-haired, at most with some scattered white ones. Females: eye hairs dense, blackish or dark brown. 5.5-7 mm. Northern and Central Europe *Heringia (Neocnemodon) pubescens* Delucchi & Pschorn-Walcher
Jizz: small, black species, wings blackish with violet sheen.
Note: if male has metatars 1 without a depression, sternite 4 flat, but sternite 3 with a prominence, see *N. fulvimanus*, 5b, which is in between *N. verrucula* and *N. pubescens*.

4.a. Males only. Sternite 3 with a keel or prominence **>** 5
4.b. Males only. Sternite 3 flat, without prominence **>** 6

123

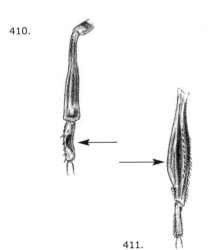

410.

411.

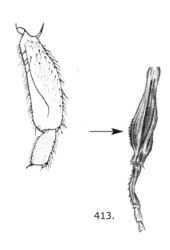

412.

413.

figure 410. *Heringia pubescens*, leg 1 of male.
figure 411. *Heringia pubescens*, leg 2 of male (Verlinden).

figure 412. *Heringia latitarsis*, metatars 1 of male (after Van der Goot, 1981).
figure 413. *Heringia latitarsis*, metatars 2 of male (Verlinden).

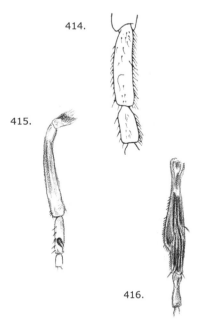

414.

415.

416.

figure 414. *Heringia fulvimanus*, metatars 1 of male (after Van der Goot, 1981).
figure 415. *Heringia vitripennis*, leg 1 of male (Verlinden).
figure 416. *Heringia vitripennis*, leg 2 of male (Verlinden).

5.a. Metatars 1 strongly broadened (figure 412); tibia 2 broadened (figure 413); sternite 2: front 1/2 with long, dense hairs; larger: 6-8 mm. Central Europe **>** *Heringia (Neocnemodon) latitarsis* Egger
5.b. Metatars 1 not broadened (figure 414); tibia 2 less broadened; sternite 2 with short hairs; smaller: 4.5-5 mm. Central Europe **>** *Heringia (Neocnemodon) fulvimanus* Zetterstedt

6.a. Males: metatars 1 and 2 without black prominence, metatars 1 with a shiny, pale pit; metatars 2 without keel (figure 415, figure 416); tibia 2: concavity without black bristles. Thoracic dorsum mainly with long white hairs, often intermixed with black ones behind the suture; face: mostly black-haired. 6-7 mm. Central Europe, in Asia to Pacific coast **>** *Heringia (Neocnemodon) vitripennis* Meigen
6.b. Males: metatars 1 and 2 with a black prominence at the back (which appears as a black dot at first sight), without pit; metatars 2 with a keel dorsally; tibia 2: concavity dressed with small black bristles (figure 417, figure 418); thoracic dorsum with white, short hairs; face: mostly white-haired. 7-8 mm. Central Europe, in Asia to Pacific coast **>** *Heringia (Neocnemodon) brevidens* Egger

417. 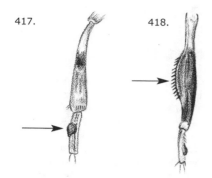 418.

figure 417. *Heringia brevidens*, leg 1 of male.
figure 418. *Heringia brevidens*, leg 2 of male
(Verlinden).

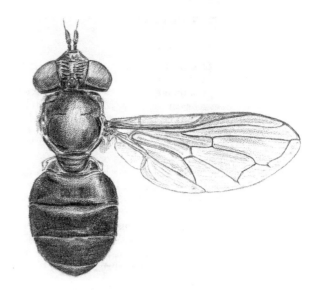

figure 419. *Lejogaster metallina*, habitus of
female (Verlinden).

LEJOGASTER

Introduction

Lejogaster is a small genus of shiny metallic
green species that occur in wet and marshy
environments. The adults can be found on
flowers (e.g. *Ranunculus* and *Peucedanum*)
and on leaves. The larvae of *Lejogaster* are
(semi-) aquatic. The tergites lack a dull
central area (unlike most *Orthonevra* and
Chrysogaster). Males have their eyes separated and the sexes must be distinguished by
the genitalia on the tip of the abdomen.

Key

1.a. Antennae completely black figure
420; legs: black, shiny metallic green;
habitus figure 419. 6-7 mm. Europe,
North Africa, in Asia to Pacific coast **>**
Lejogaster metallina Fabricius
(= *Lejogaster virgo* Rondani)
Jizz: shiny metallic green, 3rd segment of antennae large, circular.
1.b. Antennae: 3rd segment partly pale (figure 421); legs: middle segments of tarsae
pale. 5-7 mm. Europe, in Asia to Pacific
coast **>** ***Lejogaster tarsata*** Megerle in
Meigen (= *Lejogaster splendida* Meigen)
Jizz: shiny, purple-green species. Peat marshes.

420.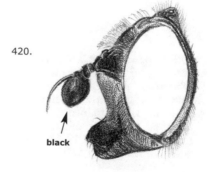

black

421.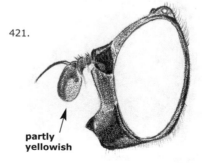

partly
yellowish

figure 420. *Lejogaster metallina*, head of
female.
figure 421. *Lejogaster tarsata*, head of
female (Verlinden).

LEJOPS

Introduction

Lejops inhabit freshwater and brackish marshes near the coast, where adults can be found on Cyperaceae such as *Scirpus maritimus* and cattail (*Typha*). Waitzbauer (1976) records that the eggs of this species are laid on the stems and leaves of emergent plants like *Typha*, and that the larvae fall into the water upon hatching. The larvae apparently remain more or less at the water surface, in association with floating plants such as *Lemna* for most of their development, but by the last instar have moved into the submerged organic ooze on the bottom of the pond or ditch where they are living.

Recognition

Longitudinal, grey stripes on the abdomen make the sole European species easy to recognize.

Key

1. Single European species. Thoracic dorsum with 5 longitudinal pale stripes of dust; abdomen elongated, tergites 2-5 with patches of grey dust which almost join to form 2 longitudinal stripes on the abdomen; tergite 2 yellow at the side margin (figure 422). 12-13 mm ❯ *Lejops vittata* Meigen

figure 422. *Lejops vittatus*, habitus of female (Verlinden).

LEJOTA

Introduction

Lejota inhabit boggy stream margins and the edges of pools in open forest from the *Picea* forest zone up into the alpine grasslands. They are black hoverflies with a broad abdomen, the face lacks a facial knob. In Sweden this species appears to be associated with open, old growth mire forest of *Betula* and *Populus tremula* (Speight, 2003). In Poland the species is found on the edges of mature mixed forest. *Lejota* is usually found close to water and the flies settle on low-growing vegetation and harvested trunks of *Picea* (Speight, 2003; T. Zeegers pers. comm.).

Key

1. Single European species. Black species with broad abdomen; antennae reddish. Males: eyes separated; sternite 4 with a small trapezoid projection. 8-10 mm ❯ *Lejota ruficornis* Zetterstedt

   Jizz: in flight like a black *Pipiza*, basks with wings open and abdomen exposed.

LEUCOZONA

Introduction

Leucozona include large hoverflies, with a white pattern on a black body. It includes the genus referred to as *Ischyrosyrphus* by other authors. *Leucozona* are damp forest species, that occur in the forest shade as well as on flowers along the forest edge. Their larvae prey on aphids. Those of *L. lucorum* (and probably *L. inopinata*) are associated with arboreal and ground aphids, those of *L. glaucia* with ground aphids and *L. laternaria* with *Caveriella* aphids on umbellifers (Rotheray, 1993).

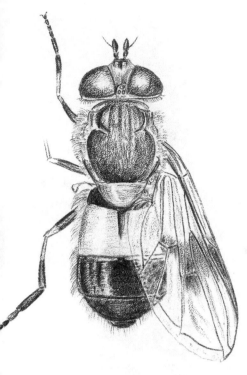

figure 423. *Leucozona lucorum*, habitus of female (Verlinden).

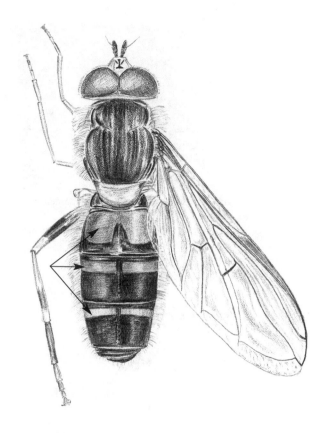

figure 424. *Leucozona glaucia*, habitus of male (Verlinden).

Recognition

Leucozona has extensive white markings on tergite 2 and small to absent markings on tergites 3 and 4. The colour of these markings is silverish-white to yellowish-white. This pattern is unique among hoverflies. The key is based on Van der Goot (1981), Verlinden (1991) and Doczal (1998).

Key

1.a. Tergite 2 mainly white, grey or pale yellow, with a black median line; tergite 3 black, often with front margin whitish, tergite 4 black; wing with a dark spot (figure 423). Exceptionally (females) abdomen black with only traces of the white on tergite 2 **>** 2
1.b. Tergite 2 with large white, grey or pale yellow spots; tergites 3 and 4 with small whitish to grey spots; wing without dark spot (figure 424) **>** 3

2.a. Tergite 4 predominantly with white hairs, black hairs may occur in the middle part; wing: more than 50% of cell bm bare; scutellum pale. 11-12 mm. Europe, in Asia to Japan **>** *Leucozona lucorum* Linnaeus
2.b. Tergite 4 with dark hairs; wing: less than 50% of cell bm bare; scutellum pale. 11-12 mm. Central Europe, Japan **>** *Leucozona inopinata* Doczkal

3.a. Scutellum yellow; tergite 2: spots often merged (figure 424, figure 425). 11-13 mm. Northern and Central Europe, in Asia east to Japan **>** *Leucozona glaucia* Linnaeus
3.b. Scutellum black, at most with pale hind border; tergite 2: spots always well-separated (figure 426, figure 427). 9-11 mm. Northern and Central Europe, in Asia east to Japan **>** *Leucozona laternaria* Müller

425.

426.

figure 425. *Leucozona glaucia* abdomen of female.
figure 426. *Leucozona laternaria*, abdomen of female (Verlinden).

figure 427. *Leucozona laternaria*, abdomen of male (Verlinden).

MALLOTA

Introduction

Mallota are large hoverflies which have a compact body, long pile or short hairs on the abdomen, and a thickened femur 3. The rat-tailed larva is saprophagous, living in standing water in cavities in decidous trees, e.g. *Acer, Aesculus, Fagus, Populus, Quercus* and *Ulmus*, usually some metres above the ground. The species apparently prefers cavities with narrow entrance holes. These holes can provide entry to central trunk cavities of considerable proportions. *Mallota* larvae have been found in these large cavities, that are partly filled with water. The larva of this species overwinters (Maibach and Goeldlin, 1989; Speight et al., 2000).

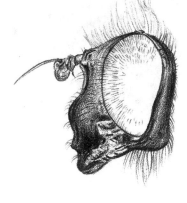

figure 428. *Mallota fuciformis*, head of male (Verlinden).

Key

1.a. Eyes bare ❯ 2
1.b. Eyes haired (figure 428); thoracic dorsum with long black hairs, scutellum and tergites 1 and 2 with long white to yellow hairs, hind margin of tergites 2 and 3 with black hairs, remaining tergites with reddish hairs. 14–17 mm. Great Britain, Central Europe, east to Iran, southwest to Spain. ❯ *Mallota fuciformis* Fabricius
Jizz: bumblebee mimic with relatively small head.

2.a. Abdomen with erect long hairs, similar to hairs on thoracic dorsum, bumblebee mimics ❯ 3
2.b. Abdomen with hairs much shorter than on thoracic dorsum, often inclined, a bee mimic (figure 429); scutellum transparent yellowish; thoracic dorsum with dense, erect yellow hairs; abdomen black, with brown hind margins of tergites. 15-17 mm. From Southern Scandinavia south to North Africa, east from Great Britain to Central Siberia ❯ *Mallota cimbiciformis* Fallén
Jizz: an oversized *E. tenax* with an elongate abdomen.

3.a. Thoracic dorsum, scutellum and abdomen with brownish to reddish-yellow hairs; tars 3 black. 11-14 mm. From Southern Scandinavia south to Germany, east to Central Siberia ❯ *Mallota megilliformis* Fallén

figure 429. *Mallota cimbiciformis*, habitus of male (Verlinden).

3.b. Thoracic dorsum and scutellum with pale grey hairs and a band of black hairs between the wings; abdomen: tergite 2 with grey hairs, hind margin of tergites 2 and 3 black-haired and remaining tergites reddish-haired; tars 3 red. 13-16 mm. Central Europe, east to Kamchatka **>** ***Mallota tricolor*** Loew
Jizz: very much like *M. fuciformis*, but hairs on thorax whitish instead of yellowish.

MELANGYNA

Introduction

Melangyna dwell in woodlands. *M. cincta* is the most abundant species of this genus in Central Europe. It prefers the edges of woodlands and shrubby vegetation, and also occurs in gardens. Other species, for example *M. ericarum* and *M. labiatarum*, are sparsely distributed throughout their range. *M. barbifrons, M. quadrimaculata* and *M. lucifera* occur early in spring and forage on blooming *Salix*. Most of the other species reach peak densities in May to June.

The larvae are oligophagous aphid predators. They seem to specialise on a few species of aphids or on aphids of a specific tree species (Rotheray, 1993).

Recognition

Melangyna species are medium-sized hoverflies, with a dark elongate abdomen. Most species have rather small spots on the abdomen which are white rather than yellow. The common *M. cincta* has yellow bands. The key is based on Speight (1988), Torp (1994) and Verlinden (1991).

Key

1.a. Tergites 3 and 4 with white to yellow spots **>** 2
1.b. Tergites 3 and 4 with yellow to orange bands; tergite 2 with a pair of yellow, triangular spots (figure 430, figure 431). 9-10 mm. Europe except extreme north, Turkey **>** ***Melangyna (Meligramma) cincta*** (= *Fagisyrphus cincta*) Fallén
Jizz: elongate with yellowish bands, similar to *Episyrphus cinctellus* and *E. auriocollis*, but tergite 2 with triangular yellowish spots.

2.a. Face: yellow, without dark stripe; tibia 1 with or without black band **>** 3
2.b. Face: with dark stripe or entirely black; tibia 1 with black band **>** 4

3.a. Antennae black; tibia and tars 1: partly dark-coloured; thoracic dorsum: 2 yellow spots just before scutellum, which may be absent in the male; tergites 3 and 4: spots rectangular (figure 432, figure 433). 8-9 mm. Northern and Central Europe, in Asia east to Pacific Ocean, Nearctic **>** ***Melangyna (Meligramma) guttata*** Fallén**>** 3
3.b. Antennae pale, in males sometimes darkened; tibia and tars 1: yellow; thoracic dorsum without yellow spots before scutellum; tergites 3 and 4: spots triangular (figure 434, figure 435). 8-10 mm. Northern and Central Europe, in Asia east to Pacific Ocean, Nearctic **>** ***Melangyna (Meligramma) triangulifera*** Zetterstedt

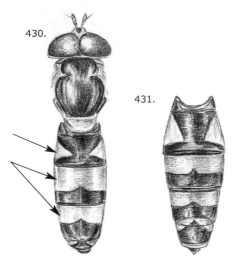

430.

431.

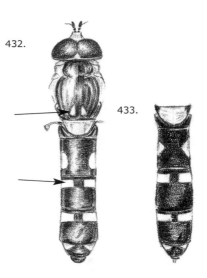

432.

433.

figure 430. *Melangyna cincta*, habitus of male.
figure 431. *Melangyna cincta*, abdomen of female (Verlinden).

figure 432. *Melangyna guttata*, habitus of male.
figure 433. *Melangyna guttata*, abdomen of female (Verlinden).

4.a. Compound eyes meeting above antennae (males) **>** 5
4.b. Compound eyes completely separate (females) **>** 16

5.a. Abdominal tergite 2 without pale marks (figure 436) **>** 6
5.b. Pair of pale marks on tergite 2 (may be very small, but nonetheless distinct) **>** 7

6.a. Eyes bare, or virtually so; face black with small yellow patches. 7-8 mm. Early spring species. Northern Europe, south to Belgium, east across Central Europe, in Asia to the Pacific Ocean **>** *Melangyna barbifrons* Fallén
6.b. Eyes densely hairy; face black; habitus figure 436. 8-9 mm. Early spring species. Northern Europe, south to Belgium, east across Central Europe, in Asia to the Pacific Ocean **>** *Melangyna quadrimaculata* Verrall

7.a. Face at least with distinct yellow patches, often yellow with black median stripe; scutellum yellow **>** 8
7.b. Face black; scutellum largely or entirely black; tergite 2 with narrow, tri-angular yellow spots, tergites 3 and 4 with pairs of narrow yellow bars that may join and which meet the lateral

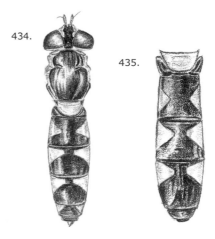

434.

435.

figure 434. *Melangyna triangulifera*, abdomen of male.
figure 435. *Melangyna triangulifera*, abdomen of male (Verlinden).

margins at full width. Elevated parts of Central Europe **>** *Melangyna (Meligramma) cingulata* Egger

8.a. Black hairs occurring across entire width of thoracic dorsum, at least over middle 1/3 of length, hairs otherwise brownish-yellow or greyish-yellow **>** 9

figure 436. *Melangyna quadrimaculata*, habitus of male (Verlinden).

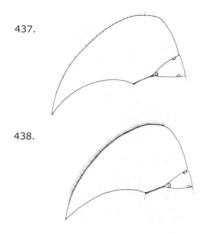

437.

438.

figure 437. *Melangyna compositarum*, eye.
figure 438. *Melangyna labiatarum*, eye (after Van der Goot, 1981).

8.b. Black hairs on thoracic dorsum confined to lateral margins, sometimes few or even absent, disc otherwise with brownish-yellow hairs ❭ 10

9.a. Wing membrane entirely covered in microtrichia; scutellar hairs no longer than the scutellum. 7-8 mm. Boreomontane: Northern Europe and mountains of Central Europe, Nearctic ❭ *Melangyna arctica* (Zetterstedt)
9.b. Radial and basal cells with areas bare of microtrichia; scutellar hairs including many longer than scutellum. 10 mm. Central Europe, Scotland. ❭ *Melangyna ericarum* Collin

10.a. Eyes haired ❭ 11
10.b. Eyes bare ❭ 15

11.a. Face: black stripe short, upwards to just over the facial knob ❭ 12
11.b. Face: black stripe long, reaching the antennal implant ❭ 13

12.a. Eyes sparsely haired (figure 437). 9-11 mm. Northern and Central Europe, south to Northern Spain, in Asia east to Pacific coast, Nearctic ❭ *Melangyna compositarum* Verrall

12.b. Eyes densely haired (figure 438). 9-11 mm. Northern and Central Europe, south to Northern Spain, in Asia east to Pacific coast, Nearctic ❭ *Melangyna labiatarum* Verrall
Note: hard to separate, may be a single species.

13.a. Pale marks on tergite 4 not distinctly triangular, not narrowing noticeably towards the mid-line; angle of approximation of eyes 90-95° ❭ 14
13.b. Pale marks on tergite 4 distinctly triangular, narrowing progressively and evenly toward the mid-line; angle of approximation of frons 110-115°. 8-10 mm. Early spring species. Northern part of Central Europe, east to Siberia ❭ *Melangyna lucifera* Nielsen

14.a. Facial prominence projecting well beyond frontal prominence; post-orbital strip wide; scutellar hairs including many yellow towards anterior margin; long hairs on general body surface black and brownish-yellow; habitus figure 439. 8-10 mm. Northern and Central Europe, found increasingly in mountainous areas further south, Siberia, Nearctic ❭ *Melangyna lasiophthalma* Zetterstedt
14.b. Facial prominence projecting no further than frontal prominence; post-orbital strip narrow; scutellar hairs black; long hairs on general body surface black and greyish-white. 8-10 mm. Northern Europe ❭ *Melangyna coei* Nielsen

figure 439. *Melangyna lasiophtalma*, habitus of male (Verlinden).

440.

441.

442.

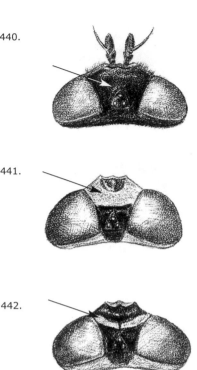

figure 440. *Melangyna barbifrons*, head of female.

figure 441. *Melangyna ericarum*, head of female (Verlinden)

figure 442. *Melangyna lasiophtalma*, head of female (Verlinden).

15.a. Wing entirely covered in microtrichia; thoracic dorsum semi-shiny with thin dust; tergites: spots yellowish ❭ 12

15.b. Wing: large areas of radial and basal cells bare of microtrichia; thoracic dorsum shiny, almost metallic blue; tergites: spots whitish. 9-11 mm. Europe, mountains of Southern Europe, Nearctic ❭ *Melangyna umbellatarum* Fabricius
Jizz: blackish with rectangular white bars on tergites.

16.a. Frons undusted and shiny, at most with rudimentary dust spots along eye margin (figure 440) ❭ 17

16.b. Frons dusted: either most or all of area from antennal insertions to ocellar triangle dusted dull (figure 441) or dust spots present which reach across 2/3 or more of the width of the frons (figure 442) ❭ 18

17.a. Tergites black, pale abdominal markings absent; eyes with short hairs ❭ *Melangyna quadrimaculata* Verrall

17.b. Pale marks present on abdominal tergites; eyes bare ❭ *Melangyna barbifrons* Fallén

18.a. Wings: cells bm and br partly bare of microtrichia ❭ 19

18.b. Wing entirely covered in microtrichia (except, occasionally, extreme base of radial cell) ❭ 21

19.a. Thoracic dorsum: humeri heavily grey-dusted, dull, contrasting sharply with brightly shining postsutural depression; frons: dust spots broad or frons entirely dusted (figure 441, figure 443) ❭ 20

19.b. Thoracic dorsum: humeri undusted, as shiny as the postsutural depression; tergites usually with yellow marks, but melanic specimens occur in which the tergites are entirely unmarked; frons: dust spots elongate (figure 442) ❭ *Melangyna lasiophthalma* Zetterstedt

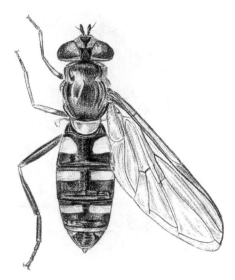

figure 443. *Melangyna umbellatarum*, habitus of female (Verlinden).

444.

445.

446.

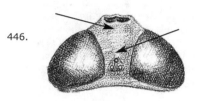

20.a. Thoracic dorsum: hairs all pale and short; scutellum yellow at sides, only a small black spot in the front corner; femora 1 and 2 with only pale hairs, posterolaterally ❯ *Melangyna umbellatarum* Fabricius

20.b. Thoracic dorsum: predominantly black-haired posterolaterally; scutellum largely black at sides; femora 1 and 2 with a mixture of black and whitish hairs on postero-lateral surface ❯ *Melangyna ericarum* Collin

21.a. Thoracic dorsum: black hairs either entirely absent or confined to lateral margins; haltere knob clear lemon yellow ❯ 22

21.b. Thoracic dorsum: hind part with scattered black hairs mixed in across almost entire width of disc; haltere knob brown/yellow-brown; head figure 444 ❯ *Melangyna arctica* Zetterstedt

22.a. No undusted, shiny band across upper part of frons, though dusting may be patchy round ocellar triangle and generally thinner on upper 1/3 of frons than on lower 2/3 of frons (figure 445, figure 446) ❯ 23

22.b. Upper part of frons black, entirely undusted, shiny, for distance wider than frontal dust-band ❯ *Melangyna coei* Nielsen

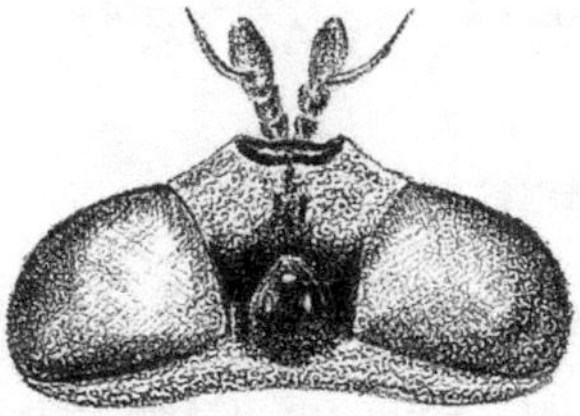

figure 444. *Melangyna arctica*, frons of female.

figure 445. *Melangyna compostarum*, head of female.

figure 446. *Melangyna labiatarum*, head of female (Verlinden).

23.a. Frons undusted just above antennae and at the vertex (elevation of ocelli) (figure 445) ❯ *Melangyna compositarum* Verrall

23.b. Frons entirely dusted, except for the lunulae just above the antennae (figure 446) ❯ *Melangyna labiatarum* Verrall

Jizz: both have dull thoracic dorsum.

Note: hard to separate, may be a single species.

MELANOGASTER

Introduction

Melanogaster are blackish, small to medium-sized hoverflies. They can be found on flowers, showing a preference for umbellifers and buttercups. The larvae are aquatic and semi-aquatic. *Melanogaster* larvae can be found in the water using their anal segments to probe for air in plant roots and shoots (see Wakkie, 2001). *Melanogaster nuda* is reported near alkaline streams, rich in organic material. *Melanogaster hirtella* is reported from peaty streams which are also rich in organic material. Finally, *Melanogaster aerosa* is reported from meso- to oligotrophic marshes where *Equisetum* species and *Hippuris vulgaris* dominate the vegetation (Maibach and Goeldlin de Tiefenau, 1994). In all species, the third larval stage hibernates; the second larval stage may also hibernate at high altitudes.

Recognition

The genus consists of black, broad-bodied flies. Males have a facial knob, females only have a protruding mouth margin. They never have two dusted longitudinal stripes on the thoracic dorsum just behind the head, as *Orthonevra* have. Tergite 1 is dull on the middle, except for *Melanogaster nuda*. The key is based on Van der Goot (1981), Verlinden (1991), Maibach et al. (1994b), Maibach and Goeldlin de Tiefenau (1995), and Vujic and Stuke (1998). *Melonogaster inornata* Loew and *M. tumescens* Loew are of uncertain status (Speight, 2003) and are not included.

Key

1.a. Male: pregenital segment (tergite 9) with erect, long hairs (figure 447). Female: thoracic dorsum with either erect hairs or with adjoining golden-yellow hairs; tergites 2-4: dull in the centre with shiny margins **>** 2
1.b. Male: pregenital segment (tergite 9) with sparse and short hairs (seems almost bare under low magnification)

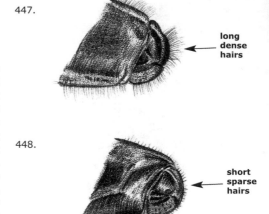

447.

long dense hairs

448.

short sparse hairs

figure 447. *Melanogaster hirtella*, tip of abdomen of male.
figure 448. *Melanogaster nuda*, tip of abdomen of male (Verlinden).

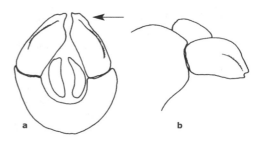

a   b

figure 449. *Melanogaster nuda*, genitalia of male (after Maibach).

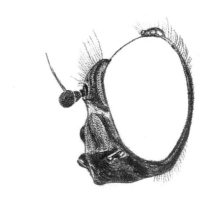

figure 450. *Melanogaster nuda*, head of male (Verlinden).

451.

452.

figure 451. *Melanogaster hirtella*, thorax of male.
figure 452. *Melanogaster curvistylus*, thorax of male (Vujic and Stuke, 1998).

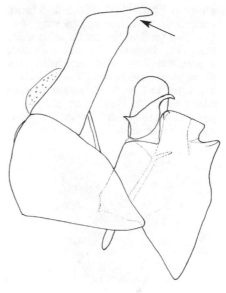

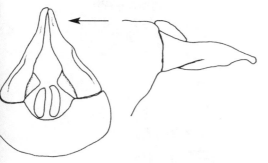

figure 455. *Melanogaster curvistylus*, genitalia of male, side view (Vujic and Stuke, 1998).

456.

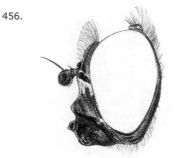

figure 453. *Melanogaster hirtella*, genitalia of male (Maibach et al., 1994).

457.

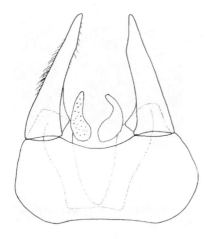

figure 454. *Melanogaster curvistylus*, genitalia of male (Vujic and Stuke, 1998).

figure 456. *Melanogaster hirtella*, head of male (Verlinden).
figure 457. *Melanogaster curvistylus*, head of male (Vujic and Stuke, 1998).

(figure 448); genitalia with short blunt surstylus (figure 449). Thoracic dorsum with short hairs (few may be long); tergites 2-4: dull in the centre with shiny margins. Female: thoracic dorsum almost bare, loosely covered with very short hairs; tergites shiny black all over. 5-6 mm. Europe **>** ***Melanogaster nuda*** Macquart (= *M. viduata* Linnaeus)

Jizz: males relatively small, dull black; females with large black patch on wing, abdomen shiny.

2.a. Male: at least most hairs on the frontal part of the thoracic dorsum yellow–brown, pale hairs often dominating; genitalia: surstylus with slender, elongated tip (figure 453). Female: unknown for *M. curvistylus*, thoracic dorsum with long, erect pale to black hairs for *M. hirtella* **>** 3

458.

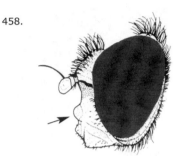

459.

460.

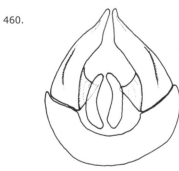

461.

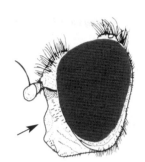

462.

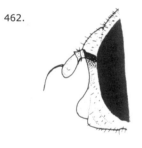

463.

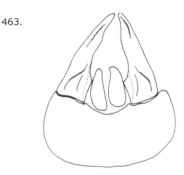

figure 458. *Melanogaster aerosa*, head of male.

figure 459. *Melanogaster aerosa*, head of female.

figure 460. *Melanogaster aerosa*, genitalia of male (Maibach et al., 1994).

figure 461. *Melanogaster parumplicata*, head of male.

figure 462. *Melanogaster parumplicata*, head of female.

figure 463. *Melanogaster parumplicata*, genitalia of male (Maibach et al., 1994).

2.b. Male: hairs on the thoracic dorsum black; eyes meet on frons on an angle of about 95°; genitalia: surstyli without elongated tip, at most pointed (figure 460). Female: thoracic dorsum with pale hairs, adpressed on the entire dorsum (*aerosa*) or erect in the anterior and adpressed in the posterior part (*parumplicata*) ❯ 4

3.a. Male: thoracic and scutellar dorsum with long and short hairs mixed, varying from completely yellowish to a large portion of dark hairs (figure 451); face broad: width of the head/width of the face under the antennae = 1.91-2.08; genitalia: surstylus with tip not curved, long and slender over 1/4 or more of its length (figure 453); head: figure 456. Female: thoracic dorsum with long, erect black to grey hairs. 6-8 mm. Central Europe ❯ *Melanogaster hirtella* Loew
Jizz: blackish species with somewhat darkened wings.

3.b. Male: thoracic and scutellar dorsum covered with short yellowish hairs of the same length (figure 452); face narrow: width of the head/width of the face under the antennae = 2.27-2.37; genitalia: surstylus with tip curved and slender over 1/6 of its length (figure 454, figure 455); head: figure 457. Female unknown. 6-8 mm. Central Europe ❯ *Melanogaster curvistylus* Vujic & Stuke
Jizz: the short pale hairs on the thoracic and scutellar dorsum distinguish this species from all other *Melanogaster* males.

4.a. Male: wing: tm reaches R4+5 at 90°; pleura: pilosity on the posterior side of the mesoanepisternum black and dense, forming a distinct fringe; head: facial knob prominent (figure 458); genitalia: figure 460. Female: thoracic dorsum: hairs (almost) adpressed, only very short hairs erect; pilosity on the bottom of tars and metatars and the innerside of the tibia of legs 1 and 3 with distinct yellow sheen; angle between face and mouth margin sharply defined (figure 459). 7-9 mm. Europe to Mongolia ❯ *Melanogaster aerosa* Loew (= *M. macquarti* auctorum)

4.b. Male: wing: tm reaches R4+5 at less than 90°; pleura: pilosity on the posterior side of the mesoanepisternum black

or white, less dense, not forming a distinct fringe; head: facial knob weak (figure 461); genitalia: figure 463. Female: thoracic dorsum: hairs erect on anterior part, adpressed on posterior part; pilosity on the tars and metatars and the inner side of the tibia of legs 1 and 3 with distinct white sheen; angle between face and mouth margin rounded (figure 462). 7-9 mm. Northern and Central Europe ❯ *Melanogaster parumplicata* Loew

MELANOSTOMA

Introduction

Melanostoma live in herbaceous vegetation, often in areas dominated by grasses. They fly close to the vegetation and are best found by sweeping a net through the vegetation. Adults visit the flowers of grasses and other plants. The larvae prey on aphids.

Recognition

Melanostoma are medium-sized hoverflies, very similar to *Platycheirus*, especially the subgenus *Pachyspira*. They differ from *Platycheirus* in their cylindrical tibia and tars 1 (male and female) and the absence of bristles on femur 1 (male). The females have triangular spots on tergites 3 and 4 (except for the black *M. dubium*), in contrast to the rectangular spots in *Platycheirus* females, which may be orange, yellow or grey.

Key

1.a. Antennae yellowish below; abdomen black with yellow spots (figure 466) (black females occur in *M. mellinum*, but these are shiny black) ❯ 2

1.b. Antennae black; tergites usually black, without yellow markings, dull in the male, (greyish) shiny in the female. 4-6 mm. Boreomontane: Northern Europe and mountains of Central Europe ❯ *Melanostoma dubium* Zetterstedt
Jizz: black species, but alpine and British males spotted.

2.a. Face and frons dusted, dull; arista short-haired (figure 464); habitus figure 466, figure 467. 8-9 mm. Europe, North Africa, in Asia east to Pacific coast, Ethiopian and Oriental regions❯ *Melanostoma scalare* Fabricius

Jizz: elongate, slender species, males: almost linear abdomen extends beyond the wings.

2.b. Face and frons shiny; arista almost bare (figure 465); habitus figure 468, figure 469. 5-7 mm. Europe, North Africa, in Asia east to Pacific coast, Nearctic ❯ *Melanostoma mellinum* Linnaeus

Jizz: more oval than *M. scalare*, abdomen does not extend beyond wings in males.

464.

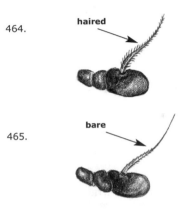

465.

figure 464. *Melanostoma scalare*, antenna.
figure 465. *Melanostoma mellinum*, antenna (Verlinden).

figure 467. *Melanostoma scalare*, abdomen of male (Verlinden).

468.

469.

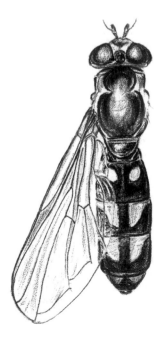

figure 466. *Melanostoma scalare*, habitus of female.

figure 468. *Melanostoma mellinum*, abdomen of female.
figure 469. *Melanostoma mellinum*, abdomen of male (Verlinden).

MERODON

Introduction

The Northwest European *Merodon* species form the vanguard of a large number of species from the Mediterranean and the steppes of Eastern Europe and beyond. In Turkey, for example, over 50 species occur (Hurkmans, 1993). The adult flies are mostly found near the ground, flying through the vegetation or resting on stones. Many species of *Merodon* appear to have a preference for Umbelliferae flowers (see Hurkmans, 1993).

The larvae all live in underground bulbs or rhizomes of monocotyledons (see Hurkmans (1993) and the references therein). *M. equestris* larvae live in bulbs and rhizomes of common garden plants such as *Amaryllis, Hyacinthus* and *Narcissus*. They are a familiar sight in the garden and pest in commercial bulb farms (Rotheray, 1993).

Recognition

Merodon are stout hoverflies with a broad abdomen. Femur 3 is typically very strong, with a large triangular projection on the underside near the tip. Large *Merodon* bumblebee mimics can be confused with other bumblebee-mimicking hoverflies. In close view femur 3 with its broad, triangular projection is characteristic. In addition, *Merodon* is often found close to the ground, while many of the other bumblebee mimics prefer higher vegetation layers. Small, more elongated *Merodon* could be confused with *Eumerus*, but *Eumerus* lack the triangular projection of femur 3. The key is based on Seguy, (1961), Van der Goot (1981), Bradescu (1991), Verlinden (1991) and Hurkmans (1993). *M. albifrons* and *M. natans* are added provisionally from material at the ZMAN in Amsterdam.

Key

1.a. Legs black, at most knees pale **>** 2
1.b. Legs: at least tibiae with extensive pale markings **>** 5

2.a. Tergites 3 and 4 with dust spots, for 1 species with blue-purple sheen; thorax and abdomen: short to medium length, erect to semi-erect hairs, hairy but not much like a bumblebee. Male: tibia 3 straight, without spurs or with a single, weakly-developed spur **>** 3
2.b. Tergites: black, without dust spots and without blue-purple sheen; thorax and abdomen with dense, long, erect hairs, almost hiding the underlying body parts, a bumblebee mimic with hairs of variable colouration. Male: tibia 3 with 2 spurs at the tip and broadened on the inner side (figure 470). 12-14 mm. Europe, North Africa, in Asia east to Japan, introduced in other parts of the world **>** *Merodon equestris* Fabricius
Jizz: large bumblebee mimic, with strong black legs and, seen from above, a relatively small head.

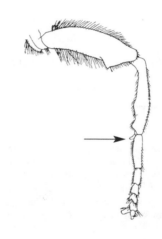

figure 470. *Merodon equestris*, leg 3 of male (after Van der Goot, 1981).

3.a. Tergite 5 black with erect hairs, similar to tergites 3 and 4; tibia 3 without spurs at the tip; thoracic dorsum with a band of black hairs between wings; larger species: 12-23 mm **>** 4
3.b. Tergite 5 reddish with semi-erect golden hairs, in contrast to the black tergites 3 and 4 with whitish and yellowish hairs; tibia 3 with a weakly-developed spur at the tip; thoracic dorsum yellow-haired, seldom with black hairs in between wings (see also 10b) smaller species: 10-12 mm. Central Europe **>** *Merodon constans* Rossi

471.

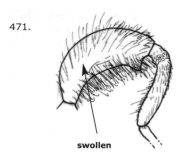

swollen

472.

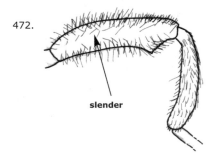

slender

473.

474.

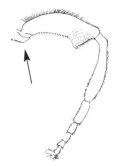

figure 471. *Merodon clavipes*, femur 3 of male.
figure 472. *Merodon aberrans*, femur 3 of male (Van Veen).

figure 473. *Merodon ruficornis*, leg 3 of male.
figure 474. *Merodon avidus*, leg 3 of male (after Van der Goot, 1981).

4.a. Femur 3 strongly curved and exceptionally enlarged (figure 471); thorax with yellowish to reddish hairs and a band of black hairs between the wings, tergite 2 with pale yellowish hairs, other tergites with more orange hairs; ground colour of tergites black without blue to purple sheen, with pale, interrupted bands. Very large species: 17-23 mm. Central and Southern Europe, Turkey ❯ *Merodon clavipes* Fallén
Jizz: very large, hairy species.

4.b. Femur 3 gently curved and not exceptionally enlarged (figure 472); thorax with brown hairs, a band of intermixed black and pale hairs in between wings, all tergites with brown hairs; ground colour of tergites black with blue-purple sheen, with pale, interrupted dust bands. Large species: 12-15 mm. Central and Southern Europe, North Africa ❯ *Merodon aberrans* Egger
Jizz: rather bare, blue-purple species.

5.a. Males, eyes meet on frons ❯ 6
5.b. Females, eyes separated ❯ 14

6.a. Trochanter 3 with a triangular or elongated protuberance (figure 473) ❯ 7
6.b. Trochanter 3 without protuberance (figure 474) ❯ 10

7.a. Tibia 3: top without protuberance (figure 475); femur 3 with or without a small protuberance on the underside, if present, located near the base ❯ 8
7.b. Tibia 3: top with spurs in the form of protuberances on the tibia; femur 3: with a protuberance on the underside, in the middle (figure 473) ❯ 9

8.a. Femur 3 without a small protuberance on the underside (do not confuse with protuberance on trochanter); thorax and abdomen with long brown hairs; tergites 3 and 4 without dust bands; smaller species: 7-10 mm. Central and Southern Europe, North Africa, Turkey ❯ *Merodon aeneus* Meigen
Jizz: small, haired species, abdomen virtually

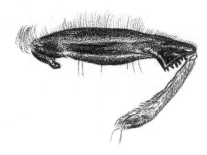

figure 475. *Merodon aeneus*, leg 3 of male
(Verlinden).

figure 476. *Merodon armipes*, leg 3 of male
(after Van der Goot, 1981).

unmarked.

8.b. Femur 3 with a small protuberance
on the underside, near the base; thorax
with short, dense, erect pile; tergites
with semi-erect to adpressed pile, tergite
2 with yellow spots, tergites 3 and 4
with greyish dust bands; larger species:
10 12 mm. Central and Southern
Europe **>** *Merodon trochantericus* Costa
Note: southern species, that extends north to
France.

9.a. Tibia 3 with 1 spur; femur 3: protuberance on the underside straight;
trochanter 3: protuberance spine-shaped
(figure 473); thorax with bronze sheen,
abdomen with weak blue sheen; tergite
2 with a pair of large yellow spots. 9-10
mm. Central and Southern Europe,
Turkey **>** *Merodon ruficornis* Meigen
9.b. Tibia 3 with 2 spurs; femur 3: protuberance curved; trochanter 3: protuberance flat, spatulate (figure 476); thorax
and abdomen black with bronze to
green sheen; tergite 2 with a pair of yellow spots. 9-11 mm. Central and
Southern Europe, Middle East, North
Africa **>** *Merodon armipes* Rondani

10.a. Tergites 3 and 4 with grey or pale
curved spots; thoracic dorsum with longitudinal pale stripes **>** 11
10.b. Tergites 3 and 4 black, without
spots; thoracic dorsum without longitudinal pale stripes (figure 477); thorax
and abdomen brownish, with bronze
sheen, covered with long pale hairs and
patches of black hairs on the middle of
tergites 2-4. 10-12 mm. Central and
Southern Europe, North Africa, in Asia
east to Kazakstan **>** *Merodon rufus*
Meigen

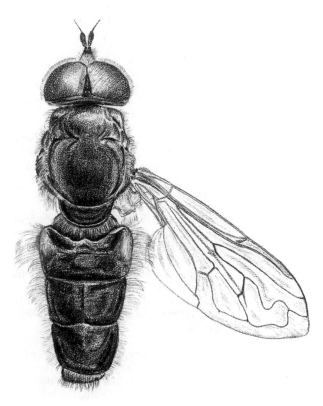

figure 477. *Merodon rufus*, habitus of male
(Verlinden).

11.a. Tergite 4 and often tergite 3 with
semi-erect to adpressed hairs; abdomen
relatively short, about twice as long as
broad; smaller species: 8-10 mm **>** 12

11.b. Tergites 3 and 4 with erect hairs, at most hairs along hind margin semi-erect; abdomen relatively long, 2.5-3 times as long as broad; larger species: 10-13 mm. Central and Southern Europe north up to Southern Scandinavia, North Africa, Turkey ❯ *Merodon avidus* Rossi

12.a. Tergite 5 with whitish or yellowish hairs, black or diffusely reddish; tibia rounded at the top, without spurs or protuberances ❯ 13
12.b. Tergite 5 reddish with long, golden hairs; tibia with a short spur at the top; thorax and abdomen black, tergites with poorly marked reddish spots, tergites 3 and 4 with grey dust spots. 10-12 mm. Central Europe ❯ *Merodon constans* Rossi
Jizz: reddish tergite 5 and long golden hairs at tip of abdomen striking.

13.a. Tergites 3 and 4 with large yellow parts, diffusely bordered with the black parts, tergites often largely yellow 8-10 mm. Southern Europe, north to Paris ❯ *Merodon albifrons* Meigen
13.b. Tergites 3 and 4 only with sharply marked yellow spots and a yellow hind margin. 7-10 mm. Southern Europe, northwards just to Paris, Middle East ❯ *Merodon natans* Fabricius
Note: both species hard to distinguish!

14.a. Thoracic dorsum with longitudinal whitish to brownish stripes of dust: at least 2 in the middle, a 3rd stripe may occur in between the middle 2 and the margins of the thoracic dorsum may be dusted ❯ 15
14.b. Thoracic dorsum without stripes of dust, margins of thoracic dorsum without dust ❯ 17

15.a. Tergite 2 with yellow spots and grey dust spots, tergites 3 and 4 black with linear yellow spots that may be dusted to almost entirely yellowish, with linear white dust spots; thoracic dorsum: dust stripes brownish ❯ 16
15.b. Tergite 2 with yellow spots, without grey spots, tergite 3 with grey spots and often with yellow spots at front margin, tergite 4 black with grey spots only, tergite 5 black (figure 478); thoracic dorsum: dust stripes white, without (5th) stripe in the middle. 10-13 mm ❯ *Merodon avidus* Rossi

16.a. Tergite 4: hind margin turning reddish or yellowish, but border with black diffuse; frons: dust spots widely separated in front of vertex. 7-10 mm. ❯ *Merodon albifrons* Meigen
16.b. Tergite 4: hind margin yellowish to reddish, with a sharp border with the black; frons: dust spots meet in front of vertex. 7-10 mm ❯ *Merodon natans* Fabricius
Note: both species hard to distinguish!

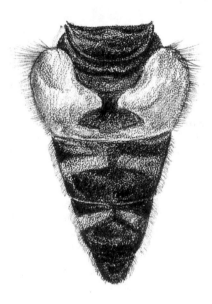

figure 478. *Merodon avidus*, abdomen of female (Verlinden).

17.a. Tergite 2 with yellowish markings, tergites 3 and 4 with or without grey to white spots; larger species; 9-12 mm ❯ 18
17.b. Tergite 2 without yellowish markings but with grey spots, tergites 3 and 4 with greyish spots (figure 479); thorax brown, with brown hairs, abdomen black, with white hairs. Smaller species: 7-9 mm ❯ *Merodon aeneus* Meigen

18.a. Tergites 3 and 4 with interrupted grey to white bands ❯ 19
18.b. Tergites 3 and 4 black, without bands; tibia and tars largely dark. 10-12 mm ❯ *Merodon rufus* Meigen

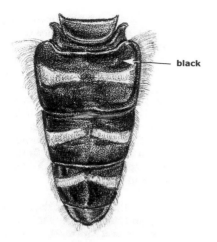

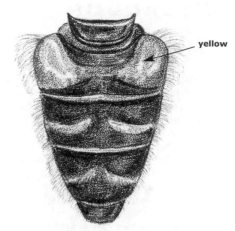

figure 479. *Merodon aeneus*, abdomen of female (Verlinden).

figure 480. *Merodon ruficornis*, abdomen of female (Verlinden).

19.b. Tergites 3 and 4: grey bands at an angle to the margins of the tergite, approaching the front margin in the middle; tergite 5 black with short white to pale yellow hairs ❯ 20

19.a. Tergites 3 and 4: grey bands straight, parallel to the margins of the tergite, tergite 5 yellowish or reddish with long golden hairs. 12 mm ❯ *Merodon constans* Rossi

20.a. Femur 3: not swollen, with straight underside; tergites 3 and 4 without yellow markings, but with grey bands (figure 480); antennae bright red ❯ 21

20.b. Femur 3 very swollen, concave on underside; antennae blackish; tergites 2-4 with extensive yellowish markings, last tergite with pale hind margin; tars 3 pale on segments 1-3, dark on segments 4 and 5. 10-12 mm ❯ *Merodon trochantericus* Costa

21.a. Frons: narrowed towards ocelli; tibia 3 flattened and broadened at the tip; tibia and tars 3 often darkened; abdomen with white to grey hairs. 9-11 mm ❯ *Merodon armipes* Rondani

21.b. Frons: hardly narrowed; tibia 3 not broadened at tip; tibia and tars 3 often extensively yellow, tarsae with last 2 segments dark; abdomen with white hairs. 9-10 mm ❯ *Merodon ruficornis* Meigen
Note: females of these 2 species are (very) hard to distinguish!

MESEMBRIUS

Introduction

Mesembrius inhabit wetlands and temporary pools in humid, seasonally-flooded, unimproved grasslands on alluvial floodplains, including slightly brackish waters (Speight, 2003). They fly in the tall, riparian vegetation (Speight, 2003), visiting the flowers of umbellifers and *Euphorbia* (Kormann, 1993).

Mesembrius is similar to *Helophilus* or a large *Parhelophilus*. Females lack most of the yellow spots on the abdomen and resemble a large *Anasimyia*. The males are immediately recognisable as their eyes meet on the frons, whereas the eyes of *Helophilus, Parhelophilus* and *Anasimyia* males are separated. In both sexes leg 3 is black, the other genera have at least extensive yellow parts on the tibia.

Key

1. Single species in Europe. Femur and tibia 3 black; tergite 2 (female) or 2 and 3 (male) with yellow spots, tergites 2-4 with grey dust band in hind part; pleurae densely covered with grey dust; face with yellow stripe. Males: eyes meet on the frons at a single point. 10-13 mm ❯ *Mesembrius peregrinus* Loew

MICRODON

Introduction

Microdon-species occur in forests, dry grasslands and bogs. They fly low in the vegetation, occasionally basking in the sun. Their larvae live in ant nests, where they feed on the eggs and larvae of ant species belonging to the genera *Formica* and *Lasius*. *M. analis* larvae occur in the mound nests of *Formica rufa*-group species and in nests of the black *Formica* and *Lasius niger*-group species, usually in rotten wood. Mature larvae, which occur near the surfaces of ant nests just prior to pupation in April, may be more easily found than adults (Speight, 2003).

Recognition

Microdon are medium-sized, broad, brownish to greenish hoverflies with long antennae and relatively small wings. This combination of characters separates them from all other hoverflies. Other hoverflies with long antennae, for example, *Chrysotoxum* or *Psarus*, are black with yellow or red markings. The key is based on Doczkal and Schmid (1999). The belief that *Microdon* may contain a number of cryptic species that can only be identified by their larvae is becoming increasingly accepted. In this key, *M. mutablis* and *M.myrmicae* form such cryptic species. Keys to larvae are provided by Doczkal and Schmid (1999) and Stubbs and Falk (2002).

Key

1.a. Pterostigma long: 3 times as long as the length of the wing margin between ends of R1 and R2+3; scutellum black, often with bronze or violet sheen (figure 481, figure 482) **>** 2

1.b. Pterostigma short: 2-2.5 times as long as the length of the wing margin between ends of R1 and R2+3; scutellum often red, but may be quite darkened (figure 483). 9-11 mm. Europe, in Asia to the Pacific coast **> *Microdon mutabilis*** Linnaeus and ***Microdon myrmicae*** Schönrogge

Note: both species can only be differentiated by larval characteristics.

2.a. Thoracic dorsum without black hairs; scutellum: hind margin between spines straight or slightly concave (figure 482); wing: cell bm with a stripe bare of microtrichia **>** 3

2.b. Thoracic dorsum posteriorly with a pair of black patches of hair or a black band of hair; hind margin of scutellum between the spines strongly concave (figure 481); wing: cell bm entirely covered with microtrichia. 9-12 mm. Europe, east to Siberia **> *Microdon devius*** Linnaeus

3.a. Antennae: 3rd segment about twice as long as 2nd; tergite 3 without dusting; tars 1: segment 2 shorter than wide; tibia 2 posteriorly with longer hairs than anteriorly, with a narrow, sharply-defined black ring; ocellular triangle often shiny; smaller species (figure 484): 7-12 mm. Europe, North Africa, in Asia east to Pacific coast **> *Microdon analis*** Macquart (= *Microdon eggeri* Mik)

481.

482.

483.

figure 481. *Microdon devius*, scutellum.
figure 482. *Microdon analis*, scutellum.
figure 483. *Microdon mutabilis*, scutellum (Verlinden).

Note: Doczkal and Schmid (1999) present *Microdon lateus* Violovitsh 1976, but they could not find satisfactory characteristics to differentiate it from *M. analis*. *M. lateus* has squarish tubercules on the scutellum instead of pointed teeth, but Doczkal and Schmid (1999) note that they found transitional forms of *M. analis* in Central Europe.

3.b. Antennae: 3rd segment 3-3.5 times as long as 2nd; tergite 3 partially dusted; tars 1: 2nd segment longer than wide; tibia 2 posteriorly with hairs as long as anteriorly, with an ill-defined dark ring; ocellular triangle dull; larger species: 11-13 mm. Europe, in Asia east to Pacific coast **>** **Microdon miki** Doczkal & Schmid (= *Microdon latifrons* auctorum)

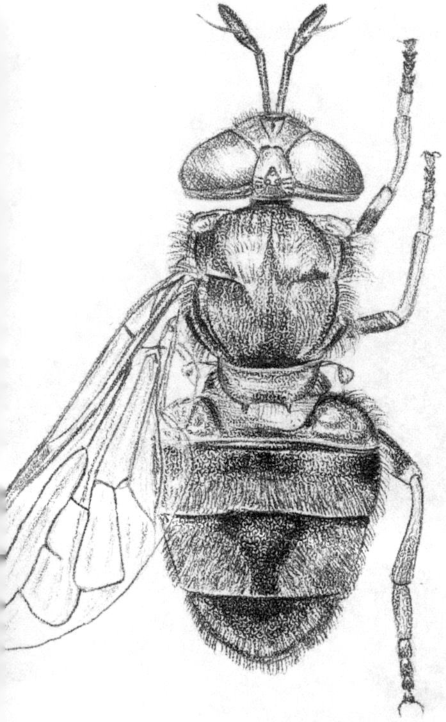

figure 484. *Microdon analis*, habitus of male (Verlinden).

MILESIA

Introduction

Milesia are very large hoverflies that mimic social wasps. The species discussed here, *M. crabroniformis*, is a convincing mimic of the hornet (as its scientific name indicates). According to Speight (2003), the buzzing sound it emits in flight is very similar to that made by the hornet. They live in deciduous forests with running water. Adults tend to dwell in the crowns of the trees, descending to feed or drink. They will follow streams out of the forest (Speight, 2003) where they visit flowers (especially flowers of *Sambucus edulis*, W. van Steenis, pers. comm.) or bask on leaves. The larvae are associated with trees, presumably living in rot holes and tree humus (Speight, 2003).

Key

1. Large, reddish-brown and yellow species; thoracic dorsum in front of wings with 2 square black patches in yellow dust, between and behind wings reddish-brown; tergites 2 and 3 with yellow patches in front, reddish-brown behind, tergite 4 yellow. 20-25 mm. Southern Europe, in France up to Bretagne **>** **Milesia crabroniformis** Fabricius

MYATHROPA

Introduction

Myathropa frequent forest edges. They can be found visiting flowers of shrubs and herbs, often agressively chasing away other hoverflies as if to defend their food source. Their larvae live in small pools of standing water, such as those remaining between tree roots or in tree cavities. They have a long anal segment, the 'rat-tail', to probe for air.

Recognition

The only Northwest European species in the genus is readily recognizable by its size, the 'skull-like' pattern of dark spots on the thoracic dorsum and the yellow and black pattern on the abdomen.

Key

1. Thoracic dorsum with a charateristic black pattern, more or less in the form of the eye cavities and the mouth of a skull (figure 485); abdomen black with yellow patterning; legs pale and black. 10-14 mm. Europe, in Asia east to Pacific coast ❯ *Myathropa florea* Linnaeus

figure 485. *Myathropa florea*, habitus (Verlinden).

MYOLEPTA

Introduction

Myolepta occur in mature forests, where they live in the canopy. Adults descend to forage and visit flowers of herbs, occasionally far from the forest. The larvae live in the rotten heartwood of deciduous trees. The key is based on Van der Goot (1981).

Key

1.a. Abdomen: tergite 2 and often tergite 3 orange with black median stripe, other tergites black (figure 486) ❯ 2
1.b. Abdomen: tergites 2 and 3 black, as are the other tergites ❯ 3

2.a. Male: face below the facial knob with a small shiny stripe, less than 1/5 of the facial width; orange spots on tergite 3 always present and reach the hind margin; black median stripe on tergite 3 less than 1/3 of the width of the tergite. Female: frons: longitudinal groove runs from the lower side of the dust spots to the front ocellus; habitus (figure 487). 10-11 mm. Central and Southern Europe ❯ *Myolepta dubia* Fabricius (= *Myolepta luteola* Gmelin)
2.b. Male: face below the facial knob with a broad shiny stripe, about 1/3 of the width of the face; orange spots on tergite 3 do not reach the hind margin, they may even be absent. Female: frons: longitudinal groove does not reach the front ocellus. 9-10 mm. Central Europe ❯ *Myolepta potens* Harris

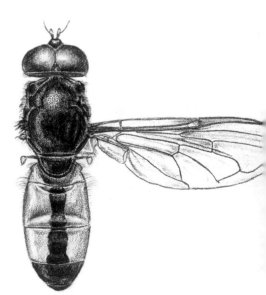

figure 486. *Myolepta dubia*, habitus of male (Verlinden).

figure 487. *Myolepta dubia*, habitus of female (Verlinden).

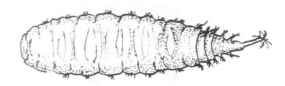

figure 488. *Neoascia tenur*, larva.

3.a. Thoracic and scutellar dorsum with short, inclined yellow to brown hairs; thoracic pleura mostly shiny black, except for anterior anepisternum; antennae pale brown. 9-12 mm. Central Europe, elevated parts of Southern Europe **›** *Myolepta vara* Panzer

3.b. Thoracic and scutellar dorsum with long, erect yellow to brown hairs; thoracic pleura dull, covered with grey dust; antennae dark brown. 9-10 mm. Central Europe **›** *Myolepta obscura* Becher

NEOASCIA

Introduction

The adults of *Neoascia* are found in damp areas near ponds and streams and in peat areas, often on the forest edge. Males can be found hovering in the shade. Males and females visit many different flowers, including *Eupatorium*, *Heracleum*, *Ranunculus* and *Taraxacum*. *Neoascia* larvae have been found in the leaf petioles of *Petasites* and in damp detritus near ponds and streams.

Recognition

Small, blackish hoverflies with a petiolate abdomen and thickened femur 3. They are often encountered in moist or wet conditions. The key is based on Barkemeyer and Claussen (1986), with *N. subchalybea* added from Bartsch et al. (2009b).

Key

1.a. Metapleurae form a continuous, sclerotinous band behind coxae 3 (figure 489), if slightly separated then antennal segment 3 about twice as long as broad and wing veins tm and tp brownish (*N. podagrica*) **›** 2

1.b. Metapleurae widely separated behind coxae 3 (figure 490), if only slightly separated or even touching at a single point then antennal segment 3 as long as broad and wing veins tm and tp clear (*N. geniculata*) **›** 5

2.a. Wing: veins tm and tp clear, without brown colouration (figure 491) **›** 3

2.b. Wing: veins tm and tp brownish (figure 492). 5-6 mm. Europe, North Africa, east to Western Siberia **›** *Neoascia podagrica* Fabricius

Jizz: common *Neoascia* with brownish tm and tp. Male: yellow band on tergite 2 almost straight.

3.a. Antenna: 3rd segment 1.5-3 times as long as wide; metapleurae broadly connected; femur 3: apex black **›** 4

3.b. Antenna: 3rd segment as long as wide; metapleurae touch in a point; femur 3: apex yellow, see 9a **›** *Neoascia geniculata* Meigen

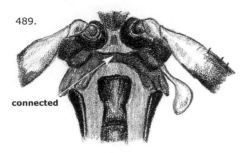

489.

connected

490.

separated

figure 489. *Neoascia podagrica*, thorax, ventral view.
figure 490. *Neoascia meticulosa*, thorax, ventral view (Verlinden).

4.a. Tergites 2 and 3: yellow band reaches the side margins of the tergite over its full width; face and mouth edge less protruding (figure 493); male: hypopygium black-haired 4-6 mm. Central Europe 〉 *Neoascia annexa* Müller (= *Neoascia floralis* Meigen)

4.b. Tergites 2 and 3: yellow band (or spots) narrowed towards the side margin, reaching the margin only over a small distance (figure 495, figure 496), or (only in female) yellow marks reduced or absent; face and mouth edge more protruding (figure 494); male: hypopygium white-haired. 3-5 mm. Europe, in Asia east to Siberia 〉 *Neoascia tenur* Harris (= *Neoascia dispar* Meigen)

Jizz: tibia 1 yellow with dark ring, metatars 1 yellow with dark patch, other segments of tars 1 yellowish. Females: abdomen not as broad as *N. meticulosa* and *N. geniculata* and seldom completely black.

5.a. Wing: veins tm and tp brownish (figure 498) 〉 6
5.b. Wing: veins tm and tp clear (figure 499) 〉 8

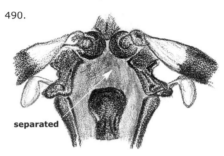

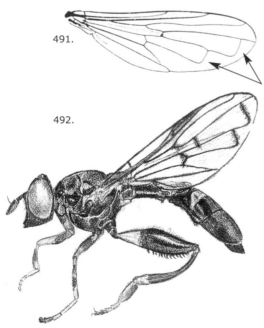

491.

492.

figure 491. *Neoascia tenur*, wing (after Van der Goot, 1981).
figure 492. *Neoascia podagrica*, habitus of female (Verlinden).

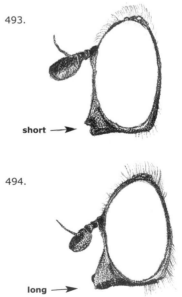

493.

short ➞

494.

long ➞

figure 493. *Neoascia annexa*, head of male.
figure 494. *Neoascia tenur*, head of male (Verlinden).

figure 496. *Neoascia tenur*, habitus of female (Verlinden).

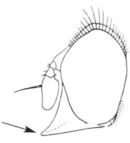

figure 497. *Neoascia petsamoensis*, head of male (after Van der Goot, 1981).

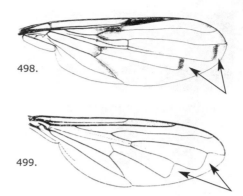

figure 498. *Neoascia interrupta*, wing.
figure 499. *Neoascia meticulosa,* wing (after Van der Goot, 1981).

6.a. Antenna: segment 3 long, 2-2.5 times as long as wide; tergite 4 black, without yellow spots ❭ 7
6.b. Antenna: segment 3 oval, at most 1.5 times as long as wide; tergite 4 with yellow spots; male: hypopygium black-haired. 5-6 mm. Northern and Central Europe, east to Western Siberia ❭ *Neoascia interrupta* Meigen
Jizz: only Neoascia with yellow spots on tergite 4.

7.a. Femur 1: yellow, at most with brownish spot; male: hypopygium black-haired; female: tergite 2 on basal 1/3 parallel-sided. 5-6 mm. Central Europe ❭ *Neoascia obliqua* Coe
Jizz: very similar to *N. podagrica.* Femur 1 yellow, yellow spot on tergite 2 in the form of a V.
7.b. Femur 1: yellow with black ring on middle; male: hypopygium white-haired; female: tergite 2 trapezoid. 5-6 mm. Central Europe ❭ *Neoascia unifasciata* Strobl
Jizz: also very similar to *N. podagrica*. Femur 1 with black ring. Only *Neoascia* with brownish tm and tp that may have tergite 2 black and tergite 3 with spots.

8.a. Antenna: 3rd segment 1.5-2 times as long as wide (figure 501); tars 1 and 2 yellow (figure 503). Male hypopygium black haired ❭ 9
8.b. Antenna: 3rd segment about as long as wide (figure 500); tibia 1 with black ring on middle, metatars and last segments of tars 1 and 2 dark (figure 502); male: hypopygium white-haired. 4-5 mm. Northern and Central Europe, east to Siberia ❭ *Neoascia geniculata* Meigen
Jizz: tibia 1 with black ring, tars 1: metatars and last 2 segments blackened, segments in between yellow; female: abdomen less broad, and often completely black.

9.a. Face: mouth edge less protruding, at most over a distance equal to 1/2 the horizontal diameter of an eye; tibia 1 and 2 yellow, at most with dark patch. Male: abdomen black with yellow spots. Female: abdomen usually black, but yellow spots may be present. 4-6 mm. Northern and Central Europe, east to Siberia ❭ *Neoascia meticulosa* Scopoli (= Neoascia aenea Meigen)
Jizz: tibia and tars 1 yellow; female with broad abdomen, usually completely black.
9.b. Face: mouth edge strongly protruding, as long as the horizontal diameter of an eye (figure 497); tibia 1 and 2 with a black ring. Males and females: abdomen black. 4-5 mm. Northern Europe, Nearctic ❭ *Neoascia subchalybea* Curran (= *Neoascia petsamoensis* Kanervo)
Jizz: black species, tibia 1 and 2 yellow with black ring, tars 1 and 2 yellow.

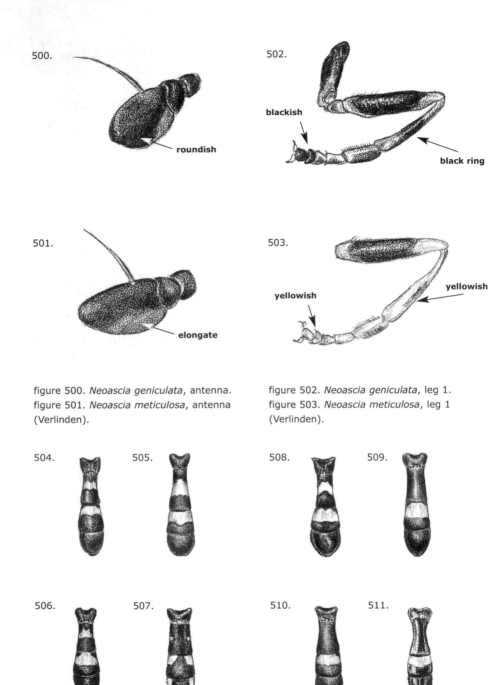

500.

roundish

502.

blackish

black ring

501.

elongate

503.

yellowish

yellowish

figure 500. *Neoascia geniculata*, antenna.
figure 501. *Neoascia meticulosa*, antenna
(Verlinden).

figure 502. *Neoascia geniculata*, leg 1.
figure 503. *Neoascia meticulosa*, leg 1
(Verlinden).

504.   505.

508.   509.

506.   507.

510.   511.

figure 504. *Neoascia podagrica*, male.
figure 505. *Neoascia annexa*, male.
figure 506. *Neoascia tenur*, male.
figure 507. *Neoascia interrupta*, male
(Verlinden).

figure 508. *Neoascia obliqua*, male.
figure 509. *Neoascia unifasciata*, male.
figure 510. *Neoascia geniculata*, male.
figure 511. *Neoascia meticulosa*, male
(Verlinden).

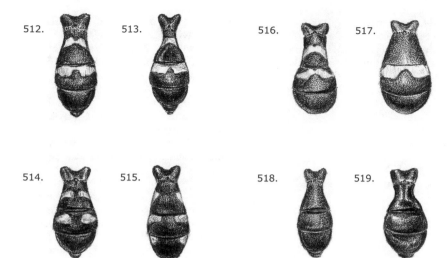

512. 513.

514. 515.

516. 517.

518. 519.

figure 512. *Neoascia podagrica*, female.
figure 513. *Neoascia annexa*, female.
figure 514. *Neoascia tenur*, female.
figure 515. *Neoascia interrupta*, female
(Verlinden).

figure 516. *Neoascia obliqua*, female.
figure 517. *Neoascia unifasciata*, female.
figure 518. *Neoascia geniculata*, female.
figure 519. *Neoascia meticulosa*, female
(Verlinden).

ORTHONEVRA

Introduction

Orthonevra are found in various damp situations, for example, in damp forests, near seepages and springs in grassland or near running water. In The Netherlands, *O. geniculata* and *O. intermedia* are typical of peatlands and bogs. Adults visit flowers of Apiaceae, Ranunculaceae and Rosaceae. Larvae are semiaquatic. The larvae of *O. nobilis* are associated with springs and flushes, where they occur in wet, organically-enriched mud. The larvae of *O. onytes* have been found among plant roots beside seepages in unmodified alpine grassland (Maibach and Goeldlin, 1994).

Recognition

Orthonevra are green metallic hoverflies, small to medium in size. Sternite 1 is shiny and typically, but not necessarily,

520.

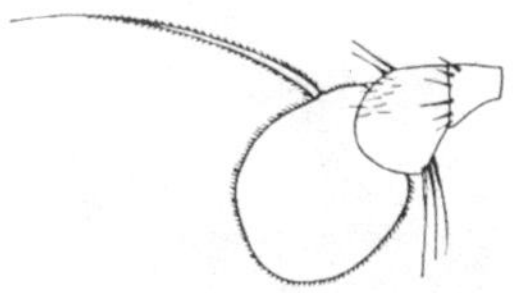

521.

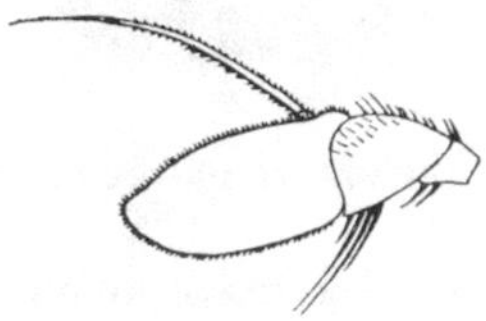

522.

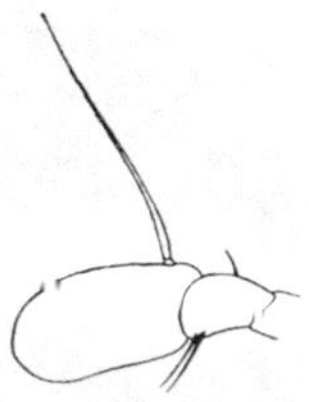

figure 520. *Orthonevra brevicornis*, antenna.
figure 521. *Orthonevra nobilis*, antenna.
figure 522. *Orthonevra plumbago*, antenna
(after Van der Goot, 1981).

Orthonevra have two longitudinal stripes of dust on the thoracic dorsum, features absent in *Chrysogaster, Melanogaster* and *Lejogaster.* The present key is based on Van der Goot (1981), Maibach et al. (1994), Seguy (1961), Sack (1932) and Speight (2003).

Key

1.a. Legs black ❯ 2
1.b. Legs bicoloured, black and yellow ❯ 7

523.

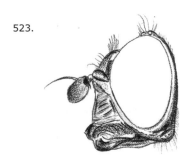

524.

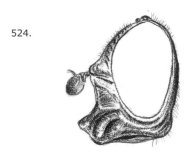

figure 523. *Orthonevra brevicornis*, head of male.
figure 524. *Orthonevra brevicornis*, head of female (Verlinden).

2.a. Antennae: 3rd segment circular, as long as wide (figure 520) ❯ 3
2.b. Antennae: 3rd segment elongated,1.5-2 times as long as wide (figure 521, figure 522) ❯ 5

3.a. Antennae: 3rd segment reddish ❯ 4
3.b. Antennae: 3rd segment black. 5-8 mm. Continental part of Central Europe ❯ *Orthonevra onytes* Seguy
Note: Possibly a synonym of *O. tristis* (Speight, 2003).

4.a. Male: frons forms an isosceles triangle, the base near the antennae longer than the part bordering the eye margin, eyes meet at an almost 90° angle; surstyli broadened on basal 1/2; female: tergite 5 without an incision at hind margin. 5-7 mm. Northern Europe, south to Northern France, east to Siberia ❯ *Orthonevra brevicornis* Loew
4.b. Male: frons forms an equilateral triangle, eyes meet at an angle smaller than 90°; surstyli not broadened on basal 1/2; female: tergite 5 with an incision at the hind margin. 7 mm. Eastern part of Central Europe ❯ *Orthonevra incisa* Loew

5.a. Antennae: 3rd segment oval, with rounded or pointed tip (figure 521) ❯ 6
5.b. Antennae: 3rd segment square, ending with a straight margin at the tip (figure 525). Eastern part of Central Europe, Southern Europe, in Asia to Pacific coast ❯ *Orthonevra frontalis* Loew

6.a. Wing: pterostigm yellow; antenna: 3rd segment rounded (figure 522). Female: tergite 5 with a deep incision at hind margin. 5-6 mm. Eastern part of Central Europe ❯ *Orthonevra plumbago* Loew

figure 525. *Orthonevra frontalis*, antenna (Verlinden).

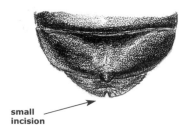

small incision

figure 526. *Orthonevra nobilis*, abdomen of female (Verlinden).

527.

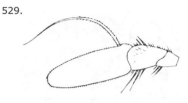

529.

528.

530.

figure 527. *Orthonevra erythrogona*, antenna.
figure 528. *Orthonevra elegans*, antenna
(after Van der Goot, 1981).

figure 529. *Orthonevra intermedia*, antenna.
figure 530. *Orthonevra geniculata*, antenna
(after Van der Goot, 1981).

6.b. Wing: pterostigm dark brown or black; antenna: 3rd segment pointed (figure 521). Female: tergite 5 with a small incision at hind margin (figure 526). 4-6 mm. Europe, in Asia to China 〉 *Orthonevra nobilis* Fallén

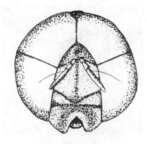

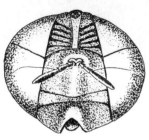

figure 531. *Orthonevra elegans*, head of male.
figure 532. *Orthonevra elegans*, head of female (Sack, 1930).

7.a. Legs: tibia pale for basal 1/3 or more, 1st segments of tarsae pale; antennae: 3rd segment 2.5 or more times longer than wide (figure 528, figure 529, figure 530) 〉 8
7.b. Legs: tibia black with only knees pale, only bases of tarsae pale; antennae: 3rd segment twice as long as it is wide (figure 527). 5-6 mm. Northern Europe, south to Northern Germany, in Asia east to Mongolia 〉 *Orthonevra erythrogona* Malm.

8.a. Eyes without dark stripe; antennae: 3rd segment about 3 times longer than wide (figure 529, figure 530) 〉 9
8.b. Eyes with dark stripe (figure 531, figure 532); antennae: 3rd segment about 4 times longer than wide (figure 528). 5-6 mm. Northern and Central Europe, in Asia east to Mongolia 〉 *Orthonevra elegans* Meigen

9.a. Wing: crossvein r-m clear, pterostigm unicolorous, yellow; face in front view smaller than width of 1 eye (figure 533, figure 534) 〉 10
9.b. Wing: crossvein r-m brown infuscate, pterostigm bicolorous: pale with dark part at base; face in front view as broad as or broader than width of 1 eye (figure 535, figure 536). 5-6 mm. Northern Europe and continental part of Central Europe, in Asia east to Mongolia 〉 *Orthonevra geniculata* Meigen

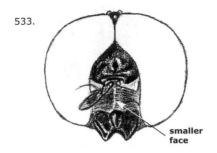

533.

smaller
face

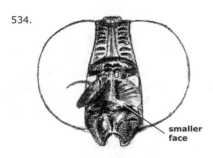

534.

smaller
face

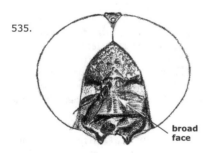

535.

broad
face

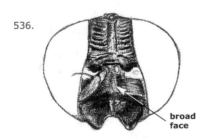

536.

broad
face

10.a. Males: genitalia: surstylus abruptly narrowed in the tip, forming a notch at the margin (figure 537a); aedeagus with a single dorsal spine at the tip (pointing to the right in figure 537b) and a large lateral spine (pointing up in figure 537b). Females: sternite 8 as figure 539, with hind corner concave, median part semi-sclerotized and bare. 6-7 mm. Central Europe, in Asia east to Pacific coast ❯ *Orthonevra intermedia* Lundbeck

10.b. Males: genitalia: surstylus gradually narrowed into the tip, without notch (figure 538a); aedeagus with 2 dorsal spines at the tip (pointing to the right in figure 538b) and a small lateral spine (pointing up in figure 538b). Females: sternite 8 as figure 540, with hind corners not concave and median part membranous and haired. 6-7 mm. Northern Europe, Siberia, in Asia east to Pacific coast ❯ *Orthonevra stackelbergi* Thompson and Torp

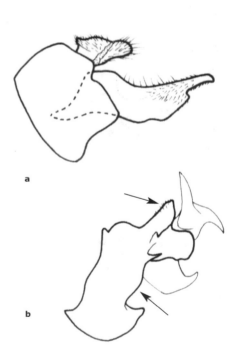

a

b

figure 533. *Orthonevra intermedia*, face of male.
figure 534. *Orthonevra intermedia*, face of female.
figure 535. *Orthonevra geniculata*, face of male.
figure 536. *Orthonevra geniculata*, face of female (Verlinden).

figure 537. *Orthonevra intermedia*, genitalia of male: (a) surstylus, (b) aedeagus (Thompson and Torp, 1982).

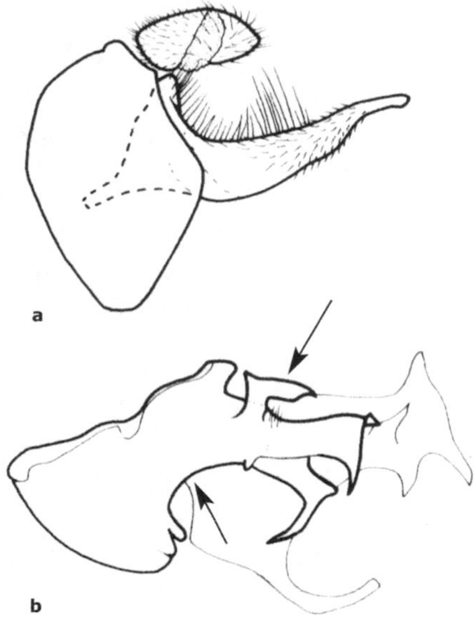

PARAGUS

Introduction

Paragus are small, thermophilous hover-flies. They are found in open areas where they dwell in open herbaceous vegetation and visit flowers such as *Senecio*. Most species prefer xerothermic conditions, but *P. finitimus* is found in marshes and peat-lands. Their larvae feed on root aphids in the ground.

Recognition

Paragus is characterized by its small size, having wings longer than its abdomen and the yellow and black pattern on its face. The abdomen can be completely black to largely red. The key is based on Goeldlin de Tiefeneau (1976), Doczkal (1996), Sommaggio (2002) and Sorokina (2002).

Key

1.a. Eyes: pilosity uniform; scutellum black. Females cannot be identified ❯ 2
1.b. Eyes: pilosity distributed in bands; scutellum black or black with yellow tip ❯ 4

2.a. Males only. Genitalia large: sternite 4 shorter than sternite 3 (figure 541); genitalia: paramere much longer than stylus (figure 544, figure 545) ❯ 3
2.b. Males only. Genitalia small: sternite 4 as long as sternite 3 (figure 542); genitalia: paramere small, hardly longer than the stylus (figure 543). 4-6 mm. Europe, Ethiopian and Nearctic regions❯ *Paragus (Pandasyophtalmus) haemorrhous* Meigen

3.a. Genitalia: parameres: without a keel on the inner side and with an upcurved upper margin, parameres typically larger, about twice the stylus length (figure 544). 6-7 mm. Central and Southern Europe, North Africa ❯ *Paragus (Pandylophtalmus) tibialis* Fallén
Note: upcurved margin of parameres well-developed, in contrast to *P. constrictus*, but visibility depends on angle of viewing.

figure 538. *Orthonevra stackelbergi*, genitalia of male: (a) surstylus, (b) aedeagus (Thompson and Torp, 1982).

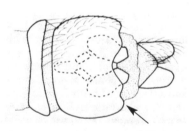

figure 539. *Orthonevra intermedium*. Sternite 8 of female (Thompson and Torp, 1982).

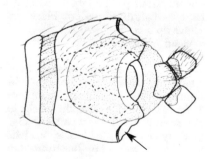

figure 540. *Orthonevra stackelbergi*. Sternite 8 of female (Thompson and Torp, 1982).

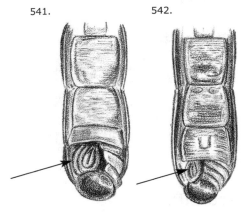

541.　542.

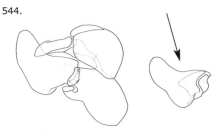

544.

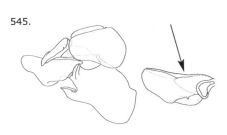

545.

figure 541. *Paragus tibialis*, last sternites of male.
figure 542. *Paragus haemorrhous*, last sternites of male (Verlinden).

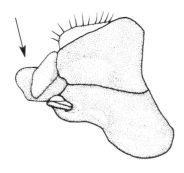

figure 543. *Paragus haemorrhous*, genitalia of male (after Van der Goot, 1981).

figure 544. *Paragus tibialis*, genitalia of male.
figure 545. *Paragus constrictus*, genitalia of male (Doczkal, 1996).

3.b. Genitalia: parameres: with a keel on the inner side along upper margin and with an almost straight upper margin, parameres typically smaller (in between *P. haemorrhous* and *P. tibialis*), about 1.5 times the stylus length (figure 545). 6-7 mm. Central and Southern Europe, Turkey ❯ *Paragus (Pandasyophtalmus) constrictus* Simic

4.a. Tergites without yellow bands but black or (partly) red, generally with silverish bands of dust (which may cover part of the red); thoracic dorsum: dull, with 2 stripes of dust. Male: genitalia: lingula elongated. Female: tergite 7 without protuberances ❯ 5

4.b. Tergites 2 and 3 with yellow bands, which may be interrupted in the middle, tergites 3-5 (also) with bands of dust; thoracic dorsum: shiny with 4-6 patches of dust (2 stripes in front part, 2 patches in hind part and 2 patches at the suture). Male: genitalia: lingula rounded. Female: tergite 7 with 2 protuberances. Central and Southern Europe, in Asia to Japan ❯ *Paragus (Paragus) quadrifasciatus* Meigen

5.a. Head in side view: face somewhat protruding, in profile largely parallel to the eye margin; scutellum black with yellow tip ❯ 6

5.b. Head in side view: face clearly protruding, in profile not following the eye margin, but extending forwards over a distance of about 1/2 the eye width (figure 546); scutellum completely black; mouth edge broadly black; tergites with black and red pattern, tergite 3 largely red with 2 black spots. Male: genitalia figure 547. Boreomontane: Northern and Central Europe, increasingly montane further south ❯ *Paragus (Paragus) punctulatus* Zetterstedt

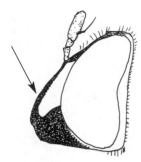

figure 546. *Paragus punctulatus*, head of male (Van Veen).

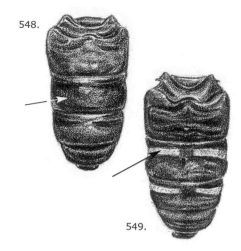

548.

549.

figure 548. *Paragus albifrons, abdomen* of male.
figure 549. *Paragus pechiolii*, abdomen of male (Verlinden).

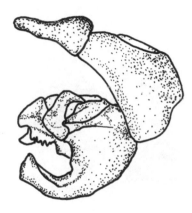

figure 547. *Paragus punctulatus*, genitalia of male (Van Veen).

6.a. Tergites 3 and 4: bands of silverish dust absent (figure 548); abdomen: tergites 3 and 4 with extensive red markings, tergites 2 and 5 black or (partially) red ❯ 7

6.b. Tergites 3 and 4: bands of silverish dust well-defined (figure 549); abdomen: entirely black or with red pattern, in the latter case usually only small red pattern on median parts of tergites 3 and 4, but *P. bicolor* with extensive red pattern on tergites 2, 3 and often 4, 5 black ❯ 9

7.a. Males: genitalia: no tooth between lingula and hypandrium. Females: tergite 5 without silverish, reflecting hairs; tergite 7 without depression ❯ 8

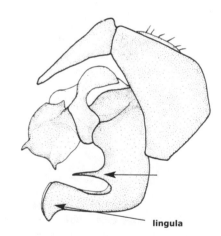

lingula

figure 550. *Paragus finitimus*, genitalia of male (after Van der Goot, 1981).

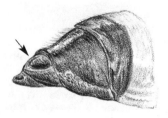

figure 551. *Paragus finitimus*, tip of female abdomen (Verlinden).

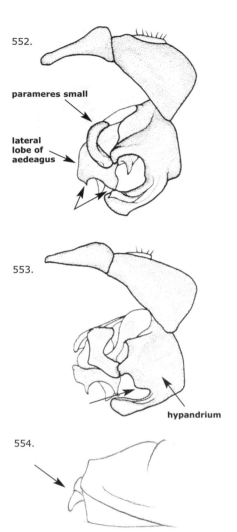

552.

parameres small

lateral lobe of aedeagus

553.

hypandrium

554.

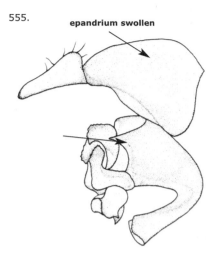

555.

epandrium swollen

556.

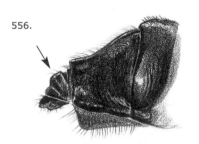

557.

parameres broad

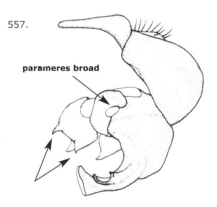

figure 552. *Paragus albifrons*, genitalia of male (after Van der Goot, 1981).
figure 553. *Paragus flammeus*, genitalia of male (after Van der Goot, 1981).
figure 554. *Paragus flammeus*, tip of female abdomen (Verlinden).

7.b. Males: genitalia: a long tooth at the border between hypandrium and lingula. Females: tergite 5 with silverish, reflecting hairs; tergites 7 and 8, in side view, of equal length; tergite 7 with a shallow depression (figure 551). 6 mm. Central Europe **❯ Paragus (Paragus) finitimus** Goeldlin

Jizz: marsh inhabitant, abdomen largely red, last segments with silverish hairs.

figure 555. *Paragus bicolor*, genitalia of male (after Van der Goot, 1981).
figure 556. *Paragus bicolor*, tip of female abdomen (Verlinden).
figure 557. *Paragus pecchiolii*, genitalia of male (after Van der Goot, 1981).

8.a. Tergite 2 black, tergites 3 and 4 black with variable red pattern (tergite 3 largely red to only a red median area), tergite 5 black. Males: lateral lobe of aedeagus with 2 teeth, the lower twice as broad as the upper, margin of hypandrium above ligula convex (figure 552). Females: tergites 7 and 8, in side view, of equal length. 6-7 mm. Central and Southern Europe, in Asia to Pacific coast ❯ *Paragus (Paragus) albifrons* Fallén

Note: very much like *P. pecchiolii*, females scarcely distinguishable except for typical specimens without dust bands.

8.b. Tergite 2 partially red, tergites 3-5 red, seldom tergite 5 black. Males: lateral lobe of aedeagus with 2 teeth of about equal size, margin of hypandrium above ligula concave (figure 553). Females: tergite 7, in side view, much smaller than tergite 8, lower margin of tergite 8 concave (figure 554). 5 mm. Central and Southern Europe, North Africa, in Asia east to Tadjikistan ❯ *Paragus flammeus* Goeldlin

9.a. Abdomen often entirely black, if red pattern occurs it does not reach the side margin. Males: genitalia: hypandrium without triangular tooth pointing towards the parameres, epandrium longer, about 1 3/4 as long as wide and slightly longer than stylus. Females: face in front view: black facial stripe broad, covering about 1/3 of the facial width and reaching the mouth edge; tergite 7 not thickened ❯ 10

9.b. Abdomen with extensive red pattern, reaching side margin at tergites 3 and 4, tergite 3 often entirely red. Males: genitalia: hypandrium upper border extended into a large triangular tooth pointing to the parameres, epandrium swollen, about 1 1/4 as long as wide and twice as long as the stylus (figure 555). Females: face in front view: black facial stripe small, covering at most 1/5 of facial width, often not reaching mouth edge; tergite 7 with a transverse swelling (figure 556). 7 mm. Central and Southern Europe, in Asia to Mongolia, Nearctic ❯ *Paragus (Paragus) bicolor* Fabricius

10.a. Tergites 2-4: bands only of silverish hairs, dust faint (or even absent, see above). Males: genitalia: parameres broad, lateral lobe with 2 teeth, the lower twice as broad as the upper, tip of lingula rounded, paramere slender (figure 552). Females: not distinguishable from next species. 6-7 mm. See 8.a. ❯ *Paragus (Paragus) albifrons* Fallén

10.b. Tergites 2-4: bands of silverish dust and hairs clearly present. Males: genitalia: parameres small, lateral lobe with 2 teeth of about equal size, tip of lingula truncated, paramere broad (figure 557). Females: indistinguishable from previous species. 6-7 mm. Central and Southern Europe, Turkey ❯ *Paragus (Paragus) pecchiolii* Rondani (= *Paragus majoranae* Rondani)

PARASYRPHUS

Introduction

Parasyrphus includes small to large flies, all of which have a black abdomen with yellow bands or spots. They can often be found visiting flowers along the forest edge. The species that appear early in the season (e.g. *P. punctulatus*) can be found on *Salix*. Later, a variety of flowers is visited. The larvae of most species prey on aphids. Larvae of *P. nigritarsis* are specialised on Chrysomelidae. They have been reared on the eggs and larvae of the Chrysomelids *Melasoma vigintipunctata* and *Melasoma aenea*, and a limited number of other Chrysomelid species.

Recognition

Most *Parasyrphus*-species are remniscent of small *Syrphus*- or *Dasysyrphus*-species. *Parasyrphus vittiger* is a small copy of *Syrphus vitripennis* in the field, but has a black facial stripe. The notable exception is *Parasyrphus nigritarsus*, which is as large as *Syrphus*- and *Epistrophe*-species and has a yellow face. The female has straight bands on the tergites, similar to *Epistrophe grossulariae* (but the latter has black antennae).

The key follows that of Theo Zeegers in the 'determinatiemap'[*], Speight (1991), and Vockeroth (1992). *P. kirgizorum* and *P. proximus* have been added provisionally from the key of Mutin, kindly translated by Theo Zeegers.

Key

1.a. Tergites 3 and 4 with spots, which may nearly join on tergite 4 ❯ 2
1.b. Tergites 3 and 4 with bands, rarely the bands on tergite 3 interrupted, but then leg 3 with much yellow ❯ 7

2.a. Face without black facial stripe, at most facial knob darkened; eyes: always with long dense hairs; boreal species ❯ 3
    Note: females cannot always be separated in this group.
2.b. Face with a black facial stripe, at least running from facial knob to mouth edge (figure 558); eyes: bare or sparsely haired, with the exception of *P. macularis* ❯ 6

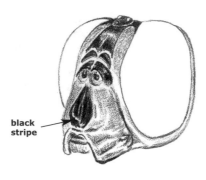

figure 558. *Parasyrphus punctulatus*, head of female (Verlinden).

black stripe

3.a. Width of face in front view, at the height of the antennae, about 2/3 the width of the head. Males: frons swollen; eyes: angle of approximation 135° or more ❯ 4
3.b. Width of face in front view, at the height of the antennae, about 1/2 the width of the head. Males: frons not swollen; eyes: angle of approximation 100° or less ❯ 5

4.a. Tergites 2-4: spots not reduced, forming yellow commas. 8-10 mm. Northern Europe. ❯ *Parasyrphus dryadis* Holmgren
    Note: status unclear, added provisionally from Van der Goot (1981).
4.b. Tergites 2-4: spots reduced to orange, often circular spots, with may be absent on 1 or more tergites. 7-8 mm. Nearctic species, occurs in Northern Scandinavia. 8-10 mm. Northern Europe, Nearctic ❯ *Parasyrphus groenlandicus* Nielsen

5.a. Antennae: 3rd segment black. Males: eyes: angle of approximation about 90°; sternites 2-4: entirely yellow, at most with narrow, dark transverse bands. 7-10 mm. Siberian boreal species ❯ *Parasyrphus kirgizorum* Peck
5.b. Antennae: 3rd segment partly pale. Males: eyes: angle of approximation about 90-100°; sternites 2-4: yellow with triangular dark spots. 7-8 mm. Northern Europe, Siberia, Nearctic ❯ *Parasyrphus tarsatus* Zetterstedt

6.a. Pterostigma light grey, in the field almost invisible; arista short, shorter than antennae, thickened over its basal 1/2 and red; eyes with short sparse hairs; abdomen: figure 559, figure 560. 6-8 mm. Northern and Central Europe, in Asia to Japan. ❯ *Parasyrphus punctulatus* Verrall
6.b. Pterostigma dark grey to black, not transparent; arista long, at least as long as the antennae, thickened over the basal 1/3 and black with at most the thickened part red; eyes with long dense hairs; abdomen: figure 561, figure 562. 8-10 mm. Northern parts of North America and Europe. 8-10 mm. Northern and Central Europe (mountains), Nearctic ❯ *Parasyrphus macularis* Zetterstedt

7.a. Antennae black ❯ 8
7.b. Antennae: 3rd segment pale below ❯ 9

8.a. Leg 3 black with at most knee yellow, apex of tibia 3 black; pterostigma dark grey; tars 1: all segments black; habitus figure 563. 8-10 mm. Northern and Central Europe, in Asia to Pacific coast, Nearctic ❯ *Parasyrphus lineolus* Zetterstedt
    Jizz: small and slender copy of a *Syrphus*, but with black facial stripe.

559.

561.

560.

562.

figure 559. *Parasyrphus punctulatus*, abdomen of male.
figure 560. *Parasyrphus punctulatus*, abdomen of female (Verlinden).

figure 561. *Parasyrphus macularis*, abdomen of male.
figure 562. *Parasyrphus macularis*, abdomen of female (Verlinden).

8.b. Tibia 3 yellow at base and apex; pterostigma light grey; tars 1: segments 4 and 5 darkened, other segments yellow. 8-10 mm. Northern and Central Europe, central Spain, east to Siberia **>** ***Parasyrphus vittiger*** Zetterstedt
Jizz: small copy of a *Syrphus*, little stouter than *P. lineolus*, with black facial stripe.

9.a. Tergites 3 and 4 without marginal sulcus; smaller: 6-9 mm; femur 3: black, at most the basal and apical 1/10 yellow; face with black stripe from mouth to facial knob (figure 564) **>** 10
9.b. Tergites 3 and 4 with marginal sulcus (figure 566); femur 3: yellow at apical 1/4 (male) or entirely yellow (face usually without black stripe, but with wide black mouth edge, figure 565); tarsae black; thoracic dorsum dulled. Female: frons: dust spots large: 9-11.5 mm. Central Europe, in Asia to Pacific coast, Nearctic **>** ***Parasyrphus nigritarsis*** Zetterstedt
Jizz: resembles *Epistrophe* with its broad yellow bands, but black tarsae contrast strongly with yellow tibiae.

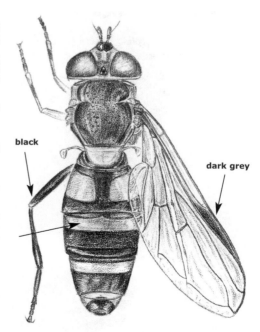

black

dark grey

figure 563. *Parasyrphus lineolus*, habitus of female (Verlinden).

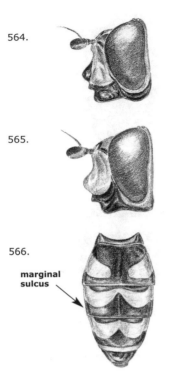

564.

565.

566.

marginal
sulcus ←

figure 564. *Parasyrphus malinellus*, head of male.

figure 565. *Parasyrphus nigritarsis*, head of male.

figure 566. *Parasyrphus nigritarsis*, abdomen of male (Verlinden).

figure 567. *Parasyrphus malinellus*, habitus of female (Verlinden).

10.a. Tars 1 and 2 darkened to black above, thoracic dorsum shiny (females may have tars 1 and 2 almost yellow); femur 3 largely or entirely black: black at base, apex may be narrowly yellow. Female: frons with small dust spots **>** 11

10.b. Tibia and tars 1 and 2 yellow; thoracic dorsum dull, greenish; femur 3 usually black with yellow base and apex (yellow base absent in 5-10% of individuals, yellow apex may also be absent). Female: frons with large dust spots. 6-8 mm. Northern and Central Europe, in Asia east to Pacific coast **>** *Parasyrphus annulatus* Zetterstedt

11.a. Sternites 3 and 4 with triangular dark spots on hind 1/2; habitus: figure 567. Males: face yellow with a black stripe; mouth edge broadly black. 8-9 mm. Northern and Central Europe, Siberia **>** *Parasyrphus malinellus* Collin

11.b. Sternites 3 and 4 with dark bands on hind 1/2. Males: face pale yellow with grey to blackish stripe; mouth edge narrowly black, occasionally pale in front. 8-9 mm. Siberia **>** *Parasyrphus proximus* Mutin

PARHELOPHILUS

Introduction

Parhelophilus are often found close to water, where they fly around vegetation. They are found resting and sunning themselves on leaves or visiting flowers such as *Iris pseudacoris* and *Potentilla palustris*. *P. frutetorum* and *P. versicolor* can also be found along forest edges. *P. consimilis* is confined to peatlands, anywhere from the vast peatlands in the west of The Netherlands to the small peaty areas alongside streams in other areas (e.g. Eifel, Germany).

The larvae live in decaying vegetation, particularly *Typha*, in ponds, fens and slow-moving streams. They are known as 'rat-tailed maggots'as they have a long anal segment that they use to reach up to the water surface for air (Rotheray, 1998).

Recognition

Parhelophilus are closely allied to *Helophilus* and share the yellowish banded thoracic dorsum with this genus. *Helophilus* is generally larger and *Helophilus* has a bare facial stripe that is lacking in *Parhelophilus*. *Lejops* and *Anasimyia* also have a banded thoracic dorsum, but the abdomen of *Lejops* is marked with white longitudinal stripes and *Anasimyia* has two black rings on tibia 3. Reemer (2000) adds a fourth, southern European species, *P. crococoronatus*, and provides a key.

568.

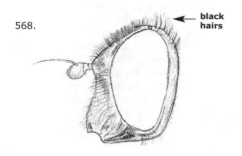

← black hairs

569.

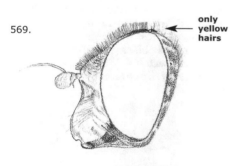

← only yellow hairs

570.

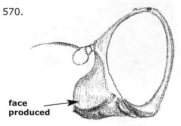

face produced →

figure 568. *Parhelophilus frutetorum*, head of female.
figure 569. *Parhelophilus versicolor*, head of male.
figure 570. *Parhelophilus consimilis*, head of male (Verlinden).

Key

1.a. Front tibiae yellow, at most with small dark patch at the tip; tergite 1 entirely grey dusted; tergite 1 with median part completely dusted; face less pronounced (figure 568, figure 569) ❯ 2

1.b. Apical 1/4 of front tibiae black, at most turning dark yellowish at the back; tergite 1 with 2 diagonal bands of dusting, leaving an undusted black median part and 2 (more or less triangular) undusted parts laterally; face in lateral view clearly protrudes beyond frons (figure 570). 8-10 mm. Northern Europe, northern part of Central Europe, Eastern Siberia ❯ ***Parhelophilus consimilis*** (Malm)
Jizz: mesotrophic peat areas, near water.

2.a. Males (eyes separated, check tip of abdomen) ❯ 3
2.b. Females (eyes separated, check tip of abdomen) ❯ 4

3.a. Femur 3: near base without posteroventral tubercle (figure 571); face swollen in profile, clearly convex; occiput with yellow hairs only; habitus: figure 573. 9-11 mm. Central and Southern Europe, east to Siberia, North Africa ❯ ***Parhelophilus versicolor*** (Fabricius)

571.

572.

tubercle

figure 571. *Parhelophilus versicolor*, femur 3 of male (Verlinden).
figure 572. *Parhelophilus frutetorum*, femur 3 of male (Verlinden).

3.b. Femur 3: near base with posteroventral tubercle, bearing a fan of bristly, black hairs (figure 572); face in profile almost flat, weakly convex; occiput dorsally has a row of black hairs among the yellow hairs. 8-10 mm. Central and Southern Europe, Siberia ❯ *Parhelophilus frutetorum* (Fabricius)

4.a. Occiput behind vertex only with yellow hairs; tergites 3 and 4 with hairs in hind part adpressed to semi-erect; face swollen in profile, convex (figure 569). 9-11 mm. ❯ *Parhelophilus versicolor* (Fabricius)
4.b. Occiput behind vertex dorsally with a row of long black hairs among the yellow hairs; tergites 3 and 4 with hairs erect on hind part; face in profile almost flat, weakly convex (figure 568). 8-10 mm. ❯ *Parhelophilus frutetorum* (Fabricius)

PELECOCERA

Introduction

Pelecocera are small hoverflies, which are usually encountered on flowers of yellow composites such as *Hieracium* and *Picris* along forest edges. Their life cycle is unknown.

Recognition

Pelecocera are small hoverflies with an elongated abdomen. Their arista is thickened and is implanted on the upper top corner of the third antennal segment, as if the third antennal segment ends in the arista. The genus is similar in appearance to *Chamaesyrphus*, but in *Chamaesyrphus* the arista is hair-like and not implanted at the end, but at three-quarters of the upper margin of the third antennal segment.

Key

1.a. Tergites 2-4 with yellow to orange spots or bands (figure 574); legs yellow; smaller: 4-5 mm. Europe, Siberia ❯ *Pelecocera tricincta* Meigen
1.b. Tergites black; larger: 7-8 mm. Eastern part of Central Europe ❯ *Pelecocera latifrons* Loew

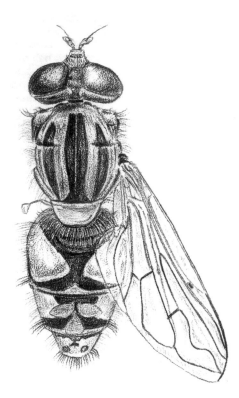

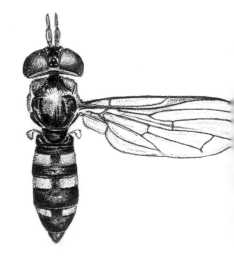

figure 573. *Parhelophilus versicolor*, habitus of male (Verlinden).

figure 574. *Pelecocera tricincta*, habitus of female (Verlinden).

PIPIZA

Introduction

Pipiza are medium-sized, blackish hover-flies with a broad abdomen. They are mostly encountered on the forest edge. *P. quadrimaculata* seems to prefer shady conditions, but other *Pipiza* regularly bask in the sun. Their larvae are predators of gall-forming aphids. *P. luteitarsis* is found in galls of *Schizoneura ulmi*, an aphid that forms galls on *Ulmus*. *P. festiva* is found on galls of the aphid *Pemphigus spyrothecae* on *Populus*. These aphids have 'soldiers', specialised aphid larvae that protect the entrance of the gall. *Pipiza* seems to overcome this barrier (Schmid, 1996).

Recognition

Pipiza contains two sets of species. The first set can be separated with confidence and contains *P. accola, P. quadrimaculata, P. luteitarsus, P. festiva, P. fasciata, P. austriaca* and *P. lugubris*. The second set contains species whose specific status is subject to doubt; it contains *P. fenestrata, P. signata, P. bimaculata, P. noctiluca* and *P. notata*. There are no satisfactory characters for *P.signata* and *P. notata*, therefore these species are omitted from the key. The key is based on Wolff (1998), an unpublished key of A. Barendregt, a translation of Violovitsh's key and information on *P. bimaculata, P. noctiluca, P. signata* and *P.fenestrata* by Th. Zeegers. D. Wolff kindly provided a pair of *P. accola* for comparison.

Key

1.a. Head in side view: implantation of antennae on upper 1/2 of head (figure 575); abdomen: more elongated, with no, 2 or 4 spots. Males: eyes: angle of approximation 85° or more. Females: frons with dust spots **>** 2

1.b. Head in side view: implantation of antennae midway on the head (figure 576); abdomen: rounded with 4 yellow spots which may be obscured (figure 577). Male: eyes: angle of approximation 80° or less. Female: frons without dust spots. 5-8 mm. Northern and Central Europe, Siberia **>** *Pipiza quadrimaculata* Panzer
Jizz: abdomen broadly oval, with 4 spots; flying in forests in shady places near running water.

2.a. Tars 1: completely yellow, rarely metatars somewhat blackened above, tars 2 and 3 largely yellow with some darkened segments **>** 3

2.b. Tars 1: at least segment 5 black, often segment 3-5 or all segments black, tars 2 and 3 largely or entirely black **>** 6

3.a. Femur 3: not thickened, without cavity below (figure 579); face in front view: broadened towards mouth edge with eye margins divergent; wing hyaline or diffusely infuscated **>** 4

3.b. Femur 3: thickened, with a cavity below where the tibia fits (figure 578); face in front view: not broadened towards mouth edge with eye margins parallel; wing: with dark spot; **>** 5

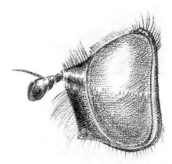

figure 575. *Pipiza lugubris*, head of male (Verlinden).

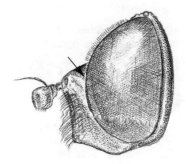

figure 576. *Pipiza quadrimaculata*, head of male (Verlinden).

figure 577. *Pipiza quadrimaculata*, habitus of female (Verlinden).

figure 580. *Pipiza festiva*, abdomen and tergite 3 of female (Verlinden).

581.

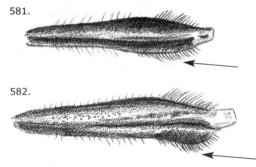

582.

figure 581. *Pipiza lugubris*, femur 3.
figure 582. *Pipiza austriaca*, femur 3 (Verlinden).

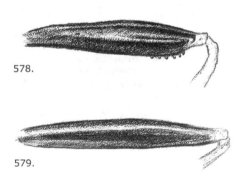

578.

579.

figure 578. *Pipiza festiva*, femur 3.
figure 579. *Pipiza luteitarsis*, femur 3 (Verlinden).

4.a. Head in front view: width of face at antennal implant about equal to the width of an eye at that height. Male: angle of approximation of eyes about 90°. Female: tergite 5 as wide as long; frons: dust spots small, each at most 1/6th of the width of the frons. 7-9 mm. Northern Europe, south to Northern France ❯ *Pipiza luteitarsis* Zetterstedt

4.b. Head in front view: face at antennal implant wider than the width of an eye at that height (figure 586, figure 587). Male: frons inflated and angle of approximation of eyes 100-110°. Female: tergite 5 twice as wide as long; frons: dust spots larger, each about 1/4th of the width of the frons. 7-9 mm. Central Europe, Siberia ❯ *Pipiza accola* Violovitsh
Jizz: a greyish *Pipiza*, covered with long, pale hairs.

5.a. Tergite 2 with large yellow spots, often broadly connected; tergite 3 black (figure 580). Males: eyes with white hairs. Female: antennae: 3rd segment 1.2 times as long as wide. 9-10 mm. Central and Southern Europe, in Asia to Pacific coast ❯ *Pipiza festiva* Meigen
Jizz: hairs on abdomen ochre-yellow.

5.b. Tergites 2 and 3 with large brown spots or bands. Males: eyes with dark hairs. Female: antennae: 3rd segment 1.5 times as long as wide. 9-10 mm. Central and Southern Europe ❯ *Pipiza fasciata* Meigen

6.a. Femur 3: club-shaped, marginally thicker than femora 1 and 2, without thickened ridges below (figure 581); thorax and abdomen pale-haired or (partly) black-haired. Males: eyes: angle of approximation 85-110° ❭ 7

6.b. Femur 3: strongly thickened, with a large ridge below (figure 582); wing with darkened spot, sharply defined at the inner and diffuse at the outer side; thorax and abdomen pale-haired. Males: eyes: angle of approximation 100-115°. Female: figure 583. 8-9 mm. Central Europe ❭ *Pipiza austriaca* Meigen

7.a. Antennae: 3rd segment 0.95-1.3 times as long as wide; wing: hyaline or with dark spot fading at outer border ❭ 8

7.b. Antennae: 3rd segment 1.3-1.5 times as long as wide (see figure 575); wing: well-marked dark spot on the middle, with a sharp outer border and fading at inner border (figure 584); thorax and abdomen with longer white and shorter black hairs. 7-9 mm. Northern and Central Europe, in Asia to Pacific coast ❭ *Pipiza lugubris* Fabricius

Jizz: dark spot in wing sharply defined at outer border.

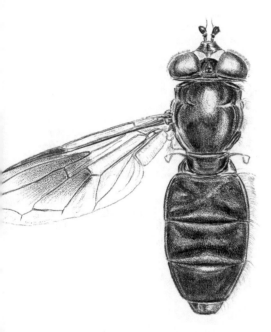

figure 583. *Pipiza austriaca*, habitus of female (Verlinden).

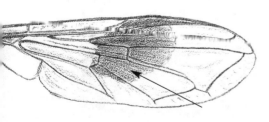

figure 584. *Pipiza lugubris*, wing (Verlinden).

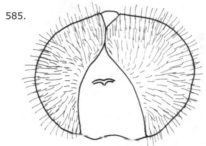

585.

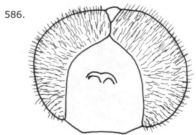

586.

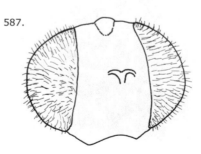

587.

figure 585. *Pipiza fenestrata*, head of male, front view.
figure 586. *Pipiza accola*, head of male, front view.
figure 587. *Pipiza accola*, head of female, front view (Van Veen).

8.a Head in front view: face narrow, width of face at antennal implant smaller than or equal to the width of an eye at that height (figure 585); face loosely dusted with a narrow shiny line below antennae; face below antennae black- or yellow-haired; thoracic dorsum completely or partly black-haired (near wings), seldom completely yellow-haired. Male: angle of approximation of eyes 85-90° **>** 9

8.b. Head in front view: face broad, at antennal implant wider than the width of an eye at that height (figure 586, figure 587); face with grey dust, without shiny line below antennae; face below antennae yellow-haired; thoracic dorsum yellow-haired; tars 1 dark above, most segments with broad black patch. Male: frons inflated and angle of approximation of eyes 100-110°. 7-9 mm. Central Europe, Siberia **>** *Pipiza accola* Violovitsh

Note: if thoracic dorsum and face predominantly yellow haired, but face smaller: compare with *P. luteitarsis* in item 4.

Jizz: a greyish *Pipiza*, covered with long, pale hairs; abdomen with small, brownish spots.

9.a. Males: face: black-haired, a few pale hairs may be present. Females: tergite 4: front half with a band of black hairs. Both sexes: tergite 3 black, seldom with a pair of spots which are a small copy of those on tergite 2 **>** 10

9.b. Males: face: predominantly white-haired. Females: tergite 4: front half without black hairs (at most a few medially). Both sexes: face in front view face not broadened towards mouth edge with eye margins parallel; tergite 2 with a pair of broad spots, tergite 3 often with a pair of elongate pale spots; tars 1 with first 3-4 segments yellow; femur 3 thickened towards tip. 9-11 mm. **>** *Pipiza fenestrata* Meigen

Note: specimens with a pale tars 1 may be confused with *P. festiva*, but that species has divergent eye margins.

Jizz: female abdomen rhomboid.

10.a. Tars 1: first 1-2 segments yellowish (figure 590); face in front view: broadened towards mouth edge with eye margins divergent; wing: darkened medianly; antenna: 3rd segment as long as wide. 6.5-10 mm **>** *Pipiza noctiluca* Linnaeus

Jizz: abdomen with small yellow spots, in many specimens even entirely black.

10.b. Tars 1: all segments black (figure 589); face in front view: not broadened towards mouth edge with eye margins parallel; wing: hyaline or diffusely infuscated; antenna: 3rd segment longer than wide. 6-8.5 mm **>** *Pipiza bimaculata* Meigen

Jizz: abdomen often with large yellow spots.

figure 588. *Pipiza fenestrata*, habitus of male (Verlinden).

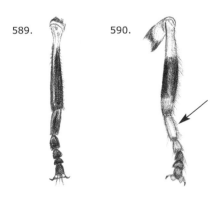

589.   590.

figure 589. *Pipiza bimaculata*, leg 1.
figure 590. *Pipiza noctiluca*, leg 1 (Verlinden).

PIPIZELLA

Introduction

Pipizella fly swiftly through the vegetation in all kinds of habitats. Of the two common species in Northwest Europe, *P. viduata* is found in a range of situations, from damp forest edges to hot chalk grasslands. *P. virens*, another relatively common species, is a thermophilous species that occurs on grasslands and at forest edges.

Recognition

Among the small, blackish hoverflies without a clear facial knob and mouth edge, *Pipizella* can be recognized by its brownish tint and the dominance of golden hairs on the thorax and legs. The third antennal segment is elongated, similar to *Heringia* sensu stricto and dissimilar to *Heringia* subgenus *Neocnemodon* and *Pipiza*.
Females cannot be identified for most species. Males can be identified by their genitalia. The key is based on Verlinden (1991, 1999).

Key

1.a. Males ❭ 2
1.b. Females ❭ not treated, only females of
  P. virens and *P. viduata* can be separated.

2.a. Eyes meet over a considerable distance. ❭ 3
2.b. Eyes narrowly separated or just touching in a single point; 3rd antennal segment little longer than wide. Sweden, Finland, Pyrenees, Northwest Spain. ❭ *Pipizella brevis* Lucas
Note: the northern populations may be named *Pipizella certa* Violovitsh, but the differences are unclear.

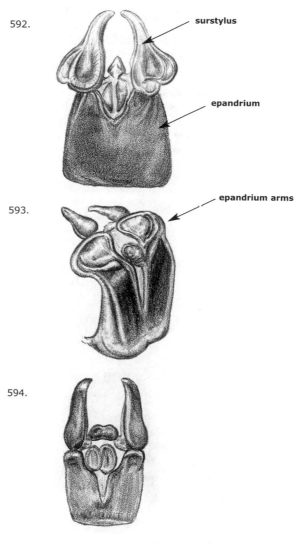

592.

593.

594.

figure 592. *Pipizella annulata*, genitalia of male.
figure 593. *Pipizella divicoi*, genitalia of male.
figure 594. *Pipizella zeneggenensis*, genitalia of male (Verlinden).

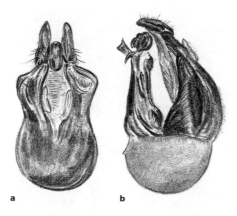

a   b

figure 591. *Pipizella pennina*, genitalia of male (Verlinden).

3.a. Sternite 3 flat; wing clear or with weak median cloud **>** 4

3.b. Sternite 3 with a transverse elevation, very convex; wing usually with a well-developed median cloud; antennae: 3rd segment quite elongate; hypopygium black-haired. 6-8 mm. Central and Southern Europe, Turkey **>** ***Pipizella maculipennis*** Meigen

4.a. Sternite 4 flat or the hind margin with a broad rounded elevation under which the genitalia are normally tucked in **>** 5

4.b. Sternite 4 with 2 lateral elevations; genital capsule voluminous; genitalia: epandrium large, in lateral view rounded posteriorly; epandrium arms broad, triangularly broadened in the middle (figure 591); hypopygium black-haired. 6-7 mm. Central Europe **>** ***Pipizella pennina*** Goeldlin

5.a. Genitalia: surstylus without rounded appendage at base, always black **>** 6

5.b. Genitalia: surstylus with the basal 1/2 broad, surpassing the epandrium arms, semicircular, exteriorly with a rounded appendage set at an angle (figure 592); surstylus often light brown or yellow, more or less translucent. 6-7 mm. Europe, Turkey **>** ***Pipizella annulata*** Macquart
Jizz: rather stout, fore and mid tibia and metatars more extensively and brighter yellow than other *Pipizella*, sharply contrasting with the black top segments of tarsi.

6.a. Genitalia: epandrium arms not flattened and widened, not strongly bent forwards; surstyli as wide as epandrium arms **>** 7

6.b. Genitalia: epandrium very large, epandrium arms flattened and widened, strongly bent forwards; surstylus very small, much narrower than epandrium arms (figure 593); genital capsule voluminous and therefore sternite 4 narrow. 6-7 mm. Central and Southern Europe, in Asia to Pacific coast **>** ***Pipizella divicoi*** Goeldlin
Jizz: sternite 3 narrow: more than 3 times as wide as its median length.

7.a. Tergites 2-4 black-haired on disk, particularly near fore and hind margins; genital capsule not so conspicuously small; surstyli shorter than epandrium **>** 8

595.

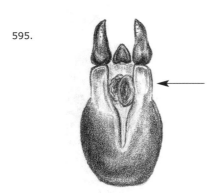

596.

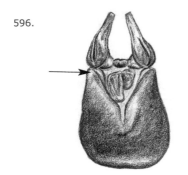

597.   598.

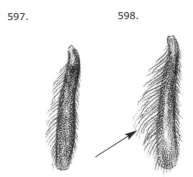

figure 595. *Pipizella viduata*, genitalia of male.
figure 596. *Pipizella virens*, genitalia of male (Verlinden).

figure 597. *Pipizella viduata*, tibia 3 of female.
figure 598. *Pipizella virens*, tibia 3 of female (Verlinden).

7.b. Tergite 4, and often 2 and 3, entirely pale-haired, rarely some black pile near its front margin; genital capsule small and flat; genitalia: surstyli as long as epandrium, in side view flattened and linear (figure 594); hypopygium pale-haired. 6-7 mm. Central Europe **〉** ***Pipizella zeneggenensis*** Goeldlin

599.

8.a. Genitalia: epandrium arms clearly present, about 1/4 of epandrium height, surstylus about 1/2 the length of the epandrium, without dorsal keel (figure 595). Male and female: hind tibia: hair fringe not so long and normally all white (figure 597); habitus figure 599. 6-7 mm. Europe, Western Siberia **〉** ***Pipizella viduata*** Linnaeus (= *Pipizella varipes* Meigen)
Jizz: smaller species.

8.b. Genitalia: epandrium arms very small or absent, epandrium approximately square, surstyli about 3/4 of the epandrium length, with a dorsal keel (figure 596). Male and female: hind tibia: with a longer hair fringe, many hairs twice as long as the maximum diameter of the tibia, often many of these hairs black and strong (figure 598); habitus figure 600. 7-8 mm. Central Europe, in Asia to Pacific coast **〉** ***Pipizella virens*** Fabricius
Jizz: larger, stouter species, reminiscent of an *Eumerus* in flight.

PLATYCHEIRUS

Introduction

Platycheirus are small to moderately large hoverflies that live in grasses and herbaceous vegetation. The abdomen is black, marked with yellow or grey spots, making the abdomen anything from black to almost yellow. Many species feed on the pollen of wind-pollinated plants, such as *Salix, Plantago, Poaceae, Cyperaceae*. They also visit flowers, for example, *Ranunculus* and *Umbelliferidae*. Many stay active during cold and rainy weather and observations suggest that lower temperatures (16 to 18 °C) are optimal for this genus. The genus contains many species with a boreomontane distribution. Some have a circumboreal distribution.

Platycheirus are rather inconspicuous in the field, both because of their small, slender and often blackish appearance and because they tend to remain hidden in

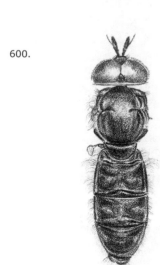

600.

figure 599. *Pipizella viduata*, habitus of male.
figure 600. *Pipizella virens*, habitus of male (Verlinden).

the vegetation. They share these features with *Melanostoma*, a closely allied genus. Catching the flies often requires sweeping a net through the vegetation. A field of *Plantago* may reveal many *Platycheirus* of several species!

The larvae are aphid predators, mostly found in the ground layer (Rotheray, 1993). Some species, such as *P. manicatus, P. peltatus* and *P. scutatus* are polyphagous, accepting many different species of aphids. Others are specialists, such as *P. fulviventris* and *P. perpallidus*. *P. fulviventris* feeds on *Hyalopterus pruni* on monocotyledonous plants in wetlands. *P. perpallidus* (and *P. immarginatus*) feed on *Trichocallis cyperi* associated with *Carex*.

Recognition

Platycheirus species all need careful identification, because there are many lookalikes and many show variation in the colour markings. A reference collection for both males and females is indispensable for resolving difficult species and individuals. Males are mainly recognised by the morphology of and the hairs and bristles on legs 1 and 2. Most species have enlarged front tibia and/or metatars in the male. Recognition of the females is more difficult and some keys practically refuse to key out many females (Van der Goot, 1981; Vockeroth, 1992) and this is not without reason. The present key aimed to include the females, but had to omit females of some species (e.g. *P. nigrofemoratus, P. groenlandicus, P. lundbecki*).

Some groups of lookalikes need particular attention in the field.

• *Platycheirus albimanus* and lookalikes. This group represents all species with greyish or silverish spots. *P. albimanus* is common in many countries and hides the other species in the field.

• *Platycheirus peltatus* and lookalikes. A group of relatively large species who are very similar in morphology; reference material is necessary to be certain of species identity. *P. peltatus* is the dominant species at lower altitudes. In The Netherlands, it can occur en masse in the Flevopolders, 3-5 metres below sea level. At higher altitudes, *P. nielseni* becomes more abundant (Van der Linden, 1991), already occurring in the Belgian Haute

Fagnes. The other species in the group are (boreo) montane species. A number of species in this group have a circumboreal distribution and are found in North America as well. Care should be taken not to overlook North American species, e.g. *P. nearcticus, P. inversus, P. octavus, P. peltatoides* (see Vockeroth, 1992). *P. peltatus* does not appear in North America, although it is common in Europe.

• *Platycheirus clypeatus* and lookalikes. Goeldlin de Tiefenau, Maibach and Speight (1990) described three further species similar to *P. clypeatus* and *P. angustatus*: *P. ramsarensis, P. occultus* and *P. europaeus*. The latter two species appear all over Europe (Van der Linden, 1991), and care is required to recognise them in the field. In addition, *P. hyperboreus* and *P. angustipes* belong to this group, but their distribution is more boreal than *clypeatus* and *angustatus*. Speight and Goeldlin de Tiefenau remark that *P. angustipes* replaces *P. occultus* and *P. europeus* in wetlands above 1000 metres in Central Europe.

• *Platycheirus manicatus* and lookalikes. Dusek and Laska (1982) review the species with a slender tibia 1 and broadened metatars 1. Some of these have grey spots, and are *P. albimanus* lookalikes. The species with yellow spots are *P. manicatus* lookalikes, such as *P. tarsalis, P. melanopsis* and *P. tatricus*. The Mediterranean *P. fasciculatus* and *P. cintoensis* also belong to this group.

• *Platycheirus scutatus* and lookalikes. Recently, three species have been described that are very similar to *P. scutatus*, two of which occur within the scope of the book: *P. splendidus* and *P. aurolateralis*.

The key is based on Van der Goot (1981), Dusek and Laska (1982), Van der Linden (1986), Speight and Goeldlin de Tiefenau (1990), Vockeroth (1992), Doczkal (1996), Stubbs (1996), Van Steenis and Goeldlin de Tiefenau (1998), Rotheray (1998), Stubbs (2002), Stubbs and Falk (2002), Doczkal et al. (2002), Smit (2003), Nielsen (2004) and Bartsch et al (2009a).

KEY

1.a. Males ❭ 2
1.b. Females (identify with care, not all species are included)-> ❭ 47

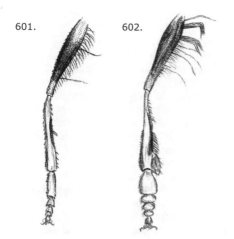

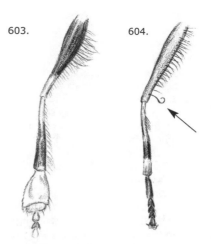

figure 601. *Platycheirus sticticus*, leg 1 of male.
figure 602. *Platycheirus albimanus*, leg 1 of male (Verlinden).

figure 603. *Platycheirus tarsalis*, leg 1 of male.
figure 604. *Platycheirus ambiguus*, leg 1 of male (Verlinden).

2.a. Tibia 1 slender, marginally increasing in width from base to apex, at apex cylindrical or flattened (figure 601) ❯ 3
2.b. Tibia 1 strongly broadened and flattened towards apex (figure 602) ❯ 22

3.a. Metatars1 slender, at least 1.7 times as long as broad measured at the maximum length and width, as wide as the tip of tibia 1 or only little wider (figure 601) ❯ 4
3.b. Metatars1 broadened, at most 1.7 times as long as wide, and distinctly broader than tip of tibia 1 (figure 603) ❯ 16

4.a. Femur 1 at the apex with a distinct, long, curled hair (figure 604); apex of tibia 1 and metatars cylindrical [note: if tibia and metatars1 are cylindrical, but femur 1 lacks the long curled hair, see *Melanostoma*!] subgenus *Pachyspyria* ❯ 5
4.b. Femur 1 without such a curled hair; apex of tibia 1 and metatars flattened ❯ 10

5.a Tergites with yellow to greyish spots, yellow spots may be heavily or lightly covered with whitish dusting ❯ 6
5.b. Tergites unspotted, black. 6-7 mm. Higher parts of Central Europe, Far East ❯ *Platycheirus immaculatus* OHara.
Jizz: small, elongate, dark species.

6.a. Femur 1 yellow, sometimes darkened behind; tergites: spots bluish, orange or yellow, sometimes covered by grey to white dusting; antennae: 3rd segment partly pale beneath or entirely brownish ❯ 7

6.b. Femur 1 black with yellow tip, subbasally usually with black bristles; tergites: spots silver-grey with thin grey dusting; antennae: 3rd segment entirely brown to black. 7 mm. Northern Europe ❯ *Platycheirus lundbecki* Collin (= *Platycheirus fjellbergi* Nielsen)

7.a. Femur 1: the long curled hair is at the end of a row of shorter but stronger hairs (figure 604); tergites: spots bluish to orange, sometimes covered by white dusting; tibia 1 laterally with a number of long, hair-like bristles which are several times longer than width of tibia ❯ 8
7.b. Femur 1: the long curled hair is at the end of a row of equally long but weaker hairs; spots on tergites yellowish with thin white dusting; tibia 1 laterally with only 1 or 2 hair-like bristles which are as long as width of tibia;. 7 mm. Northern Europe ❯ *Platycheirus transfugus* (Zetterstedt)

8.a. Haltere: knob yellow; frons with faint brown or grey dusting; tergites with blueish to orange spots with light whitish dusting ❯ 9
8.b. Haltere: knob greyish brown; frons with heavy silvergrey dusting; tergites: spots appear silvergrey because heavy greyish dust often hides yellow background completely. 7 mm. Northern and Central Europe, in Asia to Japan ❯ *Platycheirus ambiguus* Fallen

9.a. Frons covered by greyish pollinosity; femur 2 postero-laterally with rather long black hairs, the hairs near apex are longer than the thickness of the femur; metatars 3 swollen, about 1.5 times thicker than the apex of the tibia; spots on tergite bluish grey to obscure orange with light white dusting. Northern Europe and mountains of Central Europe. 7-10mm. ❯ *Platycheirus goeldlini* Nielsen

9.b. Frons covered by brownish pollinosity; femur 2 postero-laterally with rather long black hairs, which are not longer than the maximum thickness of the femur; Metatars 3 less swollen, only slightly thicker (1.2 times) than apex of the tibia; spots on tergites rather small, orange brown, covered with a light silvery-white dusting. 7-10mm. Northern Europe and Asia, mountains of Central Europe, Alaska. ❯ *Platycheirus brunnifrons* Nielsen

10.a. Femur 1 at the base without a long white hair posteriorly ❯ 11

10.b. Femur 1 at the base with a long white hair posteriorly. 5-7 mm. Northern Europe, into Asia to Eastern Siberia, Nearctic ❯ *Platycheirus aeratus* Coquillet (= *Platycheirus angustitarsis* (Kanervo), see Nielsen, 1999)

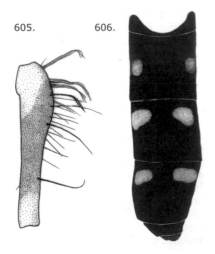

605.   606.

figure 605. *Platycheirus laskai*, 1 of male (Nielsen, 1999).
figure 606. *Platycheirus laskai*, abdomen of male (Nielsen, 1999).

11.a. Femur 1: basal 1/2 with single erect hairs or with 1 or 2 flattened hairs which are bowed at their tips; tibia 1 yellow with dark lateral stripe (*P. sticticus*) to greyish with yellow base (*P. carinatus*); tibia 2 with only short hairs or with a series of weak bristles which are strongly adpressed on at least basal 1/2 of tibia ❯ 12

11.b. Femur 1: basal 1/2 with tufts of several black, flattened, spear-like hairs which are bowed at their tips (figure 605); metatars 1 with a dark spot on apical 1/4, tarsomere 2 darkened at base; tibia 1 yellow with dark markings; tibia 2 laterally in middle with 1 or more strong, dark bristles; sternite 1 medially with many long hairs, of which the longest are several times longer than the thickness of femur 3; abdomen: figure 606. 5-7 mm. Northern Europe and mountains of Central Europe ❯ *Platycheirus laskai* Nielsen

12.a. Tibia 1 with a series of long lateral bristles ❯ 13

12.b. Tibia 1 with only scattered short, hair-like lateral bristles; tibia 1 yellow with a dark lateral marking at midlength (figure 601). 5 mm. Central Europe, into Asia to Eastern Siberia ❯ *Platycheirus sticticus* Meigen

13.a. Tars 1: at least 1st segments yellow; femur 1: no large, black bristles basoventrally ❯ 14

13.b. Tars 1 black; femur 1: with 2-4 strong, black bristles baso-ventrally; tibia 1 black with at most a narrow yellow base; tibia 1: bristles short at the base, longer towards apex. The last 3 or 4 bristles are very close together and almost forming a fascicule. 5-8 mm. Northern Europe, Siberia, Nearctic ❯ *Platycheirus carinatus* Curran (probably a synonym of *Platycheirus hirtipes* Kanervo, see Nielsen, 1999)

14.a. Metatars 3 distinctly thickened, thicker than tibia; anterior angle of approximation of eyes broad (about 103-115°); pale spots on tergite 2 absent, at most vaguely marked; lower side of metatars 1 with a darkened band. 6-7.5 mm. Northern Europe, Siberia, Nearctic ❯ *Platycheirus subordinatus* Becker

14.b. Metatars 3 marginally thickened, as thick as tibia; anterior angle of approximation of eyes less broad (about 90-100°); pale spots on tergite 2 as clear as those on tergites 3 and 4; lower side of metatars 1 with distinct dark spot on outer part of apical 1/2 **>** 15

15.a. Metatars 1 rather narrow, distinctly narrower than 1/2 its length; outer part of lower side of metatars 1 with dark spot occupying about 1/3 of length, inner part without broad dark spot; legs paler, middle leg with about basal 1/2 of tibia and about apical 1/4 of femur pale. 7-8 mm. Northern Europe, into Asia to Japan **> *Platycheirus latimanus* Wahlberg**

15.b. Metatars 1 somewhat broader, almost as broad as 1/2 its length; outer part of lower side of metatars 1 with distinct dark spot occupying usually less than 1/3 length, inner part usually with somewhat darkened extensive spot occupying nearly inner 1/2 of segment and longer than 1/2 of length; legs darker, middle leg with about basal 1/3 only of tibia and apex of femur pale. 7-9 mm. Central Europe, into Asia to Japan **> *Platycheirus complicatus* Becker**

Note: *P. kittilaensis* (16a) is very similar, it has a more triangular metatarsus, 2nd tarsal segment broader than long (slightly variable): ratio maximum width:length = 1.44. *P. complicatus* has a narrower (less triangular) metatarsus, 2nd tarsal segment squarish, as broad as long: ratio maximum width:length = 1.08.

figure 607. *Platycheirus discimanus*, leg 1 of male (Verlinden).

16.a. Markings on tergites silverish or whitish **>** 17
16.b. Markings on tergites yellow **>** 18

17.a. Tarsus 2: first 2 segments yellow, the 1st strongly compressed, the 2nd slightly compressed, last 3 segments black; leg 1: figure 607. 7-8 mm. Central Europe, into Asia to Pacific coast, Nearctic **> *Platycheirus discimanus* Loew**
17.b. Tarsus 2: all segments black and rounded. 7-8 mm. Northern Europe, Siberia, Nearctic **> *Platycheirus groenlandicus* Curran** (= *Platycheirus boreomontanus* Nielsen)

18.a. Whole outer side of tibia 1 with row of about 10-15 distinct dark hairs. 7-9 mm. Northern Europe **> *Platycheirus kittilaensis* Dusek and Laska**

Note: *P. complicatus* (13b) is very similar, it has a narrower (less triangular) metatarsus, 2nd tarsal segment squarish, as broad as long: ratio maximum width:length = 1.08. *P. kittelaensis* has a more triangular metatarsus, 2nd tarsal segment broader than long (slightly variable): ratio maximum width:length = 1.44.

18.b. Outer side of tibia 1 without complete row of distinct dark hairs **>** 19

19.a. Mouth edge projecting, but not beyond facial knob (e.g. figure 608); abdomen: spots whitish to pale yellow, smaller than *manicatus* (e.g. figure 609); thoracic dorsum shiny, at most a little prunosity in the middle **>** 20
19.b. Mouth edge projecting beyond facial knob (figure 610); abdomen with 4 pairs of spots; abdomen broad, spots yellow, large (figure 611); thoracic dorsum dull; leg 1: figure 612. 9-10 mm. Europe, North Africa, in Asia to Siberia, Nearctic **> *Platycheirus manicatus* Meigen**

20.a. Segment 3 of tarsus 1 abruptly narrower than segments 1 and 2, segment 3 as wide as segments 4 and 5 **>** 21
20.b. Segments 1 and 2 of tarsus 1 taper into segment 3, segment 3 broader than segments 4 and 5 (figure 613); abdomen with 3 pairs of reddish spots, those on tergite 2 small. 7-8 mm. Northern Europe, mountains of Central Europe, into Asia to Eastern Siberia **> *Platycheirus melanopsis* Loew**

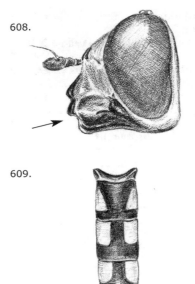

608.

609.

figure 608. *Platycheirus tarsalis*, head of male.
figure 609. *Platycheirus tarsalis*, abdomen of male (Verlinden).

610.

611.

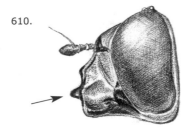

figure 610. *Platycheirus manicatus*, head of male.
figure 611. *Platycheirus manicatus*, abdomen of male (Verlinden).

612.

613.

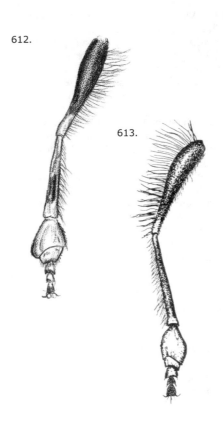

figure 612. *Platycheirus manicatus*, leg 1 of male (Verlinden).
figure 613. *Platycheirus melanopsis*, leg 1 of male (Sack, 1930).

21.a. Outer side of tibia 3 with tuft of long black hairs in the middle; anterior angle of approximation of eyes usually narrower than 100°; hairs on face pale or at least predominantly pale; metatars 1 broader: length:width = 6:5 (use reference material). 8-9 mm. Central and Southern Europe, into Asia to Eastern Siberia ❯ *Platycheirus tarsalis* Schummel

21.b. Outer side of tibia 3 with dispersed short black hairs only; anterior angle of approximation of eyes distinctly broader than 100°; hairs on face dark except for some pale hairs on mouth edge; metatars 1 longer: length:width = 5:3 (use reference material). 8-9 mm. Central Europe ❯ *Platycheirus tatricus* Dusek and Laska

614.

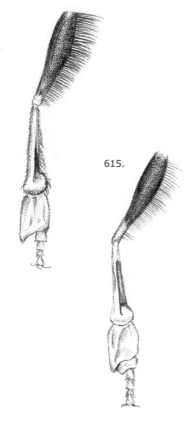
615.

figure 616. *Platycheirus holarcticus*, leg 1 of male (Vockeroth, 1992).

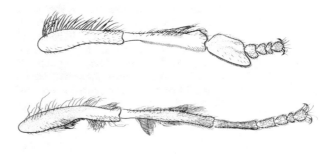

figure 617. *Platycheirus amplus*, legs 1, 2 of male (Vockeroth, 1992).

figure 614. *Platycheirus peltatus*, leg 1 of male.

figure 615. *Platycheirus parmatus*, leg 1 of male (Verlinden).

22.a. Posterior surface of femur 1 densely covered with rather strong, nearly uniform slightly flattened black hairs (figure 614); femur 2 with a deep concavity anteriorly; antennae: 3rd segment partly yellow ❯ 23

22.b. Posterior surface of femur 1 without flattened black hairs (figure 615); femur 2 without concavity or with a shallow concavity (for *P. jaerensis*); antennae black or 3rd segment partly yellow (only for *P. parmatus, P. jaerensis*) ❯ 26

23.a. Metatars1 with a moderately strong dorsal keel on apical 1/2, but without distinct keel on basal 1/2; wing with bare area in basal cells or completely covered with microtrichia ❯ 24

23.b. Metatars1 with distinct dorsal keel over its entire length; thoracic dorsum and scutellum with hairs mostly black; wing completely covered with microtrichia. 7-10 mm. Northern Europe, Nearctic presumably Siberia❯ *Platycheirus holarcticus* Vockeroth

24.a. Tibia 2 ventrally constricted in apical 1/2, constriction preceded by a swelling implanted with dense black hairs and ending in a weakly to strongly swollen apex, tibia 2 without tuft of long black hairs at 1/3 of tibia length; larger species: 8-10 mm ❯ 25

24.b. Tibia 2 not ventrally constricted, without black-haired swelling halfway and tip not swollen (only a little notched), at 1/3 of tibia length an anterior tuft of dense long black hairs (figure 617); wing: cell bm with a median bare area in basal 1/4. 7-9 mm. Northern Europe, elevated parts of Central Europe, Nearctic presumably Siberia ❯ *Platycheirus amplus* Curran

25.a Constriction on tibia 2 shallow and about 1/2 the length of tibia, preceded by a swelling and ending in weakly swollen apex, this apex only with short erect hairs; tibia 2 often yellow; wing:

cell bm with a median bare area in basal 1/2. 8-10 mm. Northern and Central Europe, into Asia to Pacific coast ❯ *Platycheirus peltatus* (Meigen)

25.b Constriction on tibia 2 profound and about 1/3 the length of tibia, preceded by a swelling and ending in a strongly swollen apex, upper part of swollen apex with a tuft of longish hairs, directed upwards parallel to the shaft (figure 618); tibia 2 often darkened; wing: cell bm without bare area, although less trichose at base. 8-10 mm. Northern Europe, elevated parts of Central Europe, into Asia to Pacific coast, Nearctic ❯ *Platycheirus nielseni* Vockeroth

26.a. Femur 1: 2 striking tufts of black hair near base (figure 619) ❯ 27
26.b. Femur 1: without hair tufts near base ❯ 32

27.a. Trochanter 2 with slender, finger-like process; tarsus 1: segment 2 short, about 1/6 the length of segment 1 (figure 619) ❯ 28
27.b. Trochanter 2 without ventral process; tarsus 1: segment 2 relatively long, about 2/5 the length of segment 1 ❯ 30

28.a. Face: in front view, dull black or with golden dust above antennae and silverish dust below antennae; wing: basal cells completely covered in microtrichia; tibia 2 in side view bent in middle or near apex and ventrally with some very long hairs; tergite 4 square or nearly so ❯ 29
28.b. Face: in front view, dust completely silverish, also above antennae; wing: basal cells partly free of microtrichia at base; tibia 2 in side view bent 1/3 from apex and laterally with only short hairs (figure 621); tergite 4 elongate, about 1.25 times as long as wide (figure 620). 7-10 mm. Europe, in Asia east to Pacific coast, Nearctic ❯ *Platycheirus scutatus* Meigen

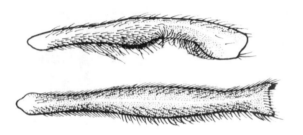

figure 618. *Platycheirus nielseni*, male: tibia of leg 2 (upper) and leg 3 (lower)(Vockeroth, 1992).

figure 619. *Platycheirus scutatus*, leg 1 of male (Verlinden).

figure 620. *Platycheirus scutatus*, abdomen of male (Verlinden).

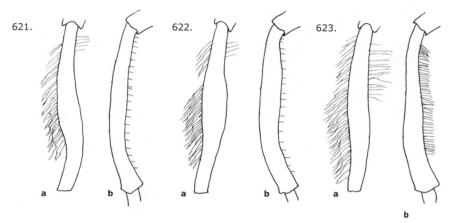

figure 621. *Platycheirus scutatus*, tibia 2: (a) dorsal view, (b) lateral view.
figure 622. *Platycheirus splendidus*, tibia 2: (a) dorsal view, (b) lateral view.
figure 623. *Platycheirus aurolateralis*, tibia 2: (a) dorsal view, (b) lateral view (Smit, 2003).

figure 624. *Platycheirus albimanus*, leg 1 of male (Verlinden).

figure 625. *Platycheirus nigrofemoratus*, leg 1 of male (Vockeroth, 1992).

29.a. Tergites 2-4 square, tergite 2 with distinct spots; tars 1 with segment 4 about twice as wide as long and highly asymmetrical; tibia 2 in lateral view: bent in the middle, rather swollen in basal 1/2, ventrally with short hairs only, in dorsal view the long hairs separated in 2 groups, base and tip bare (figure 622). 6-8 mm. Presumably Northern and Central Europe **>** *Platycheirus splendidus* Rotheray

29.b. Tergites 2-4 slightly elongate, tergite 2 with spots often lacking, at most vaguely marked; tars 1 with segment 4 more than twice as wide as long and reasonably symmetrical; tibia 2 in lateral view: bent near apex, narrowest in basal 1/2, ventrally with equally distributed long whitish hairs (longer than diameter of tibia), medially superimposed by long hairs, apical 1/3 with erect hairs often well-developed (figure 623). 6-8 mm. Described in 2002, only known from Britain at present **>** *Platycheirus aurolateralis* Stubbs

30.a. Tibia 1 uniformly broadened on basal 3/4 and then more strongly broadened posteriorly (figure 624).); **>** 31

30.b. Tibia 1 uniformly broadened from base to apex, its margins slightly divergent throughout (figure 625); face: in front view, with golden dust above antennae and silverish dust below antennae; wing: cell bm with at most a small bare area near base; 6-8 mm. Northern Europe, Siberia, Nearctic **>** *Platycheirus nigrofemoratus* Kanervo
Note: if tibia 1 and metatarsus 1 slender, metatars about twice as long as broad, abdomen black with small greyish spots, see *Platycheirus laskai*, 11.b.

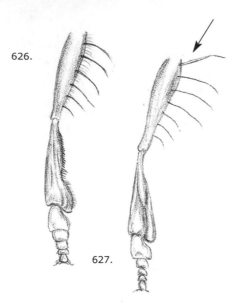

626.

627.

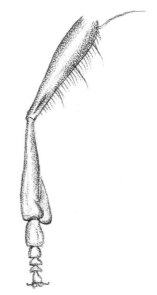

figure 626. *Platycheirus scambus*, leg 1 of male.
figure 627. *Platycheirus immarginatus*, leg 1 of male (Verlinden).

figure 630. *Platycheirus angustatus*, leg 1 of male (Verlinden).

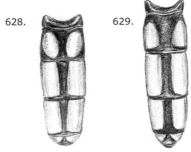

628.   629.

figure 628. *Platycheirus immarginatus*, abdomen of male.
figure 629. *Platycheirus scambus*, leg 1 of male (Verlinden).

31.a. Wing: cell bm with a small to rather large bare area; tibia 1: the extension at the posterior part of the top triangular, it ends acute; metatars 1: posterior edge gently curved; face: in front view, dust completely silverish, also above antennae. 7-10 mm. Europe, in Asia east to Pacific coast, Nearctic **>** *Platycheirus albimanus* Fabricius (= *P. cyaneus* (Muller)
31.b. Wing: cell bm entirely trichose; tibia 1: the extension at the posterior part of the top almost rectangular, it ends truncate; metatars 1: posterior edge angular at 1/3rd from the top. 6-9 mm. Northern Europe and Asia. **>** *Platycheirus urawakensis* (Matsumura)

32.a. Femur 1: posterior surface with 3-6 long, moderately strong, evenly spaced, black bristles (figure 626, 627); femur 2: apical 1/2 of anteroventral surface with nearly regular row of 7-16 short stout bristles, usually mostly black (in rare cases all yellow) **>** 33
32.b. Femur 1: posterior surface with uniform fine hairs or with 1 row of 4-5 long weak black bristles; femur 2: anteroventral surface without row of short stout bristles **>** 34

33.a Femur 1: subbasally with tuft of 2-3 long white hairs and dorsally with 6 long black bristles (figure 627); tibia 1 gradually increasing in width, without incision. 8 mm. Northern Europe and Atlantic parts of Central Europe, Nearctic **>** *Platycheirus immarginatus* Zetterstedt
33.b Femur 1: without long white hairs at base and dorsally with 5 long black bristles (figure 626); tibia 1 irregularly increasing in width, with a small incision at the outer margin. 8-9 mm. Northern and Central Europe, in Asia to Pacific coast, Nearctic **>** *Platycheirus scambus* Staeger

figure 631. *Platycheirus podagratus*, leg 1 of male (Sack, 1930).

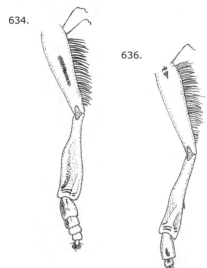

figure 634. *Platycheirus fulviventris*, leg 1 of male.

figure 636. *Platycheirus perpallidus*, leg 1 of male (Van der Linden).

632.   633.

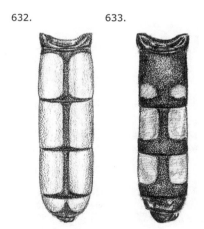

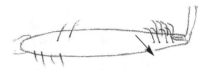

figure 635. *Platycheirus perpallidus*, femur 2 of male, from below (Verlinden).

figure 632. *Platycheirus fulviventris*, abdomen of male.

figure 633. *Platycheirus clypeatus*, abdomen of male (Verlinden).

34.a. Femur 1: posterior surface with sub-basal tuft of 2-3 closely adpressed long white or yellowish hairs with wavy apices (figure 630); antennae black ❯ 35

34.b. Femur 1: without subbasal tuft of white hairs, either with uniform fine hairs or with well-spaced longer hairs or bristles over most of its length; antennae black or partly yellow ❯ 45

35.a. Tibia 1 uniformly broadened from base to apex; leg 3 yellow or yellow with black markings ❯ 36

35.b. Tibia 1 strongly and abruptly broadened on apical 1/3, slightly narrower apically than preapically; leg 3 black with yellow knee; abdomen slender. 8 mm. Northern Europe and elevated parts of Central Europe, east to Central Siberia, Nearctic ❯ *Platycheirus podagratus* (Zetterstedt) (= *Platycheirus nudipes* Becker, see Nielsen, 1999)
Jizz: much like *Melanostoma scalare* on first sight.

36.a. Abdomen with 4 pairs of spots, spots on tergite 5 sometimes merged into 1 (figure 632) ❯ 37

36.b. Abdomen with 3 pairs of spots, tergite 5 black, at most somewhat paler at front margin (figure 633) ❯ 38

637.

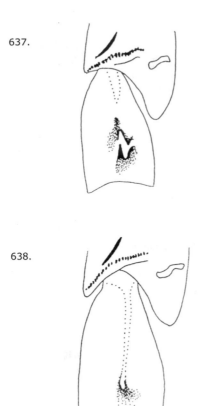

638.

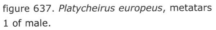

639.

640.

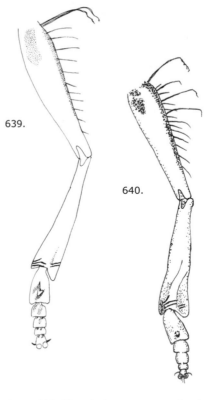

figure 639. *Platycheirus europaeus*, leg 1 of male.

figure 640. *Platycheirus angustatus*, leg 1 of male (Van der Linden).

figure 637. *Platycheirus europeus*, metatars 1 of male.

figure 638. *Platycheirus clypeatus*, metatars 1 of male (Speight).

37.a. Femur 2 without a series of bent hairs ventrally near the tip; femur 1 with dense black hairs; tibia 1 abruptly broadened at mid-length, then more or less parallel-sided to the tip (figure 634). 8-9 mm. Central and Southern Europe, in Asia east to Pacific coast ❯ *Platycheirus fulviventris* Macquart

37.b. Femur 2 near tip with a series of anteroventral black hairs that are strongly bent towards the base of the femur (figure 635); femur 1 with dense yellow hairs, intermixed with some black ones; tibia 1 uniformly broadened from base to apex (figure 636). 8-9 mm. Northern Europe and northern part of Central Europe, in Asia east to Pacific coast ❯ *Platycheirus perpallidus* Verrall

38.a. Wing: cell bm with a bare spot at base; metatars 1 with v-shaped incision (figure 637) ❯ 39

38.b. Wing completetely trichose; metatars 1 without v-shaped incision (figure 638)❯ 40

39.a. Notopleurae and anepisternum dusted, semi-shiny; spots on tergite 3 at most 2/3 length of tergite; femur 1: long black hairs uniformly bent (figure 639). 7-8 mm. Europe ❯ *Platycheirus europaeus* Goeldlin, Maibach and Speight

39.b. Notopleurae and anepisternum shiny; spots on tergite 3 at least 4/5 length of tergite; femur 1: some long black hairs suddenly bent at tip (figure 640). 6-8 mm. Northern and Central Europe, in Asia east to Pacific coast, Nearctic ❯ *Platycheirus angustatus* (Zetterstedt)

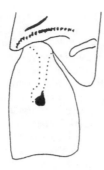

figure 641. *Platycheirus occultus*, metatars 1 of male (Speight).

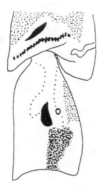

figure 642. *Platycheirus angustipes*, metatars 1 of male (Speight).

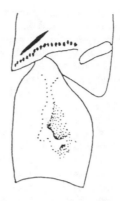

figure 643. *Platycheirus hyperboreus*, metatars 1 of male (Speight).

40.a. Metatars 1: ventral surface with, close to its longitudinal axis, a bare, pale, shiny streak confined to the basal half, ending in a small pale, shiny, rounded depression containing a black mark (figure 641) ❭ 41

40.b. Metatars 1: ventral surface with, close to its longitudinal axis, a bare, shiny area that reaches the apical half, often in the form of a streak reaching the apical 1/4, without rounded depression with black marking on basal half (figure 638) ❭ 42

41.a. Tars 1: apical half of all segments without dark brown/black blotches ventrally (figure 641); femur 1 black/dark brown for less than half its length; surstyli entirely pale-haired. 7-8 mm. Europe, increasingly montane further south ❭ *Platycheirus occultus* Goeldlin, Maibach and Speight

41.b. Tars 1: all segments with a large brown/black blotch ventrally, in distal half of segment (figure 642); femur 1 black/dark brown for 3/4 of length; base of surstyli often with long, black, bristly hairs mixed with the pale yellow bristly hairs. 7-8 mm. Elevated parts of Central Europe ❭ *Platycheirus angustipes* Goeldlin

Jizz: legs very dark for the group including *P. clypeatus*.

42.a. Metatars 1: ventrally with a broad pit in the middle, without much dark coloration; tars 3 black or brownish, little or no contrast between central and last segments; femur 3 black at base; tibia 3 almost completely dark, leaving small pale top and base; tergites with yellow spots undusted or heavily greyish dusted ❭ 43

42.b. Metatars 1: ventrally with a black marked, narrow pit in the top part; tars 3 with central 2 segments pale and last 2 segments black; femur 3 pale at base; tibia 3 with broad pale top and base; tergites with yellow spots undusted ❭ 44

43.a. Tergites with yellow spots without grey dusting. Larger: 7-9 mm. Northern Europa above polar circle, Siberia. ❭ *Platycheirus magadanensis* Mutin

Note: provisionally included.

43.b. Tergites with spots heavily dusted, yellow colour hardly visible. Smal: 5-7 mm. Northern Europe, in Asia to Pacific coast, Nearctic. ❭ *Platycheirus hyperboreus* Staeger

44.a. Femur 1: posterior hairs extending to near apex; tars 1 and 2: orange; face with eye margins almost parallel from level of antennal insertions to lower margin of facial knob (figure 644). 7-8 mm. Northern and Central Europe, east to Siberia, Nearctic ❯ *Platycheirus clypeatus* (Meigen)

44.b. Femur 1: posterior hairs absent in apical 1/2; tars 1 and 2: at least apical segments dark; face usually broadening progressively from level of antennal insertions to lower margin of facial knob (figure 645). 7-8 mm. Atlantic part of Northern Europe ❯ *Platycheirus ramsarensis* Goeldlin, Maibach and Speight

45.a. Tergites 3 and 4 with yellow spots; antennae: segment 3 yellow beneath ❯ 46

45.b. Tergites 3 and 4 with silvery spots, without trace of yellow markings; antennae black. 7-9 mm. Northern Europe, Nearctic ❯ *Platycheirus varipes* Curran (= *Platycheirus argentatus* Ringdahl)

46.a. Tibia 1 with many posterior hairs longer than tibial width; segment 2 of tarsus 1 much wider than long and only slightly narrower than metatars; femur 2: anteriorly without concavity, anteroventrally with only long hairs. 8-10 mm. Northern and Central Europe, east to Siberia, Nearctic ❯ *Platycheirus parmatus* Rondani (= *P. ovalis* (Becker))

46.b. Tibia 1: posterior hairs much shorter than tibial width; segment 2 of tarsus 1 slightly longer than wide, much narrower than metatars; femur 2 has a shallow concavity anteriorly beyond midlength, bordered below by short dense slightly curved black bristles, otherwise with only short anteroventral hairs. 8-10 mm. Northern Europe and mountains of Central Europe, Nearctic ❯ *Platycheirus jaerensis* Nielsen

47.a. Abdominal spots greyish to blueish or completely lacking ❯ 48

47.b. Abdominal spots yellow to whitish ❯ 59
Note: take care, not all species are included and the range of variation is not always known; for example, melanistic females of normally spotted species occur, al be it rarely.

48.a. Spots on tergites 3 and 4 not joined or absent ❯ 49

48.b. Spots on tergites 3 and 4 joined, with an extension to the hind border of the tergite (figure 648) ❯ *Platycheirus ambiguus* Fallén

49.a. Abdomen black, without spots ❯ 50

49.b. Abdomen with spots, sometimes reduced ❯ 52

644.

645.

figure 644. *Platycheirus clypeatus*, face of male.
figure 645. *Platycheirus ramsarensis*, face of male (Speight).

figure 646. *Platycheirus parmatus*, leg 1 of male (Van der Linden).

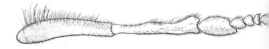

figure 647. *Platycheirus jaerensis*, leg 1 of male (Vockeroth, 1992).

50.a. Thoracic dorsum: hairs pale ❯ 51

50.b. Thoracic dorsum: hairs mainly black; tergites shining black, in certain angles of light with barely discernable spots ❯ *Platycheirus carinatus* Curran

51.a. Antennae black; wing: with bare parts ; femur 1: largely black with pale top ❯ *Platycheirus subordinatus* Becker

Note: *P. laskai* may lack spots, has brown to black antennae, but femur 1 pale.

51.b. Antennae pale below; wing: completely trichose; femur 1: largely pale. ❯ *Platycheirus immaculatus* ÔHara

52.a. Femur 1 and 2: at least basal half black ❯ 53

52.b. Femur 1 and 2: almost entirely yellow ❯ 57

53.a Abdomen with 4 pairs of spots because tergite 5 is spotted (figure 649); antennae: 3rd segment as long as wide; ❯ 54

53.b Abdomen with 3 pairs of spots, tergite 5 black ❯ 55

Note: *P. carinatus* may have faint dust spots, 3rd segment of antennae as long as wide, tergite 5 black, femur 1 almost entirely black.

54.a. Head from above: vertex at the smallest point narrower than an eye width; frons shining black; sternite 2: hairs short, with a few exceptions ❯ *Platycheirus discimanus*

54.b. Head from above: vertex at the smallest point wider than an eye width; frons with dust band; sternite 2: many long hairs present, especially in front ❯ *Platycheirus groenlandicus*

figure 648. *Platycheirus ambiguus*, abdomen of female (Verlinden).

figure 649. *Platycheirus discimanus*, abdomen of female (Verlinden).

55.a. Posterior anepisternum: entirely covered with grey dusting; Scutellum: without bristles or with pale bristles only ❯ 56

55.b. Posterior anepisternum: lower part shining, undusted; Scutellum: some of the long bristles black ❯ *Platycheirus urawakensis*

56.a. Face narrower than eye width; scutellum with pale bristles only ❯ *Platycheirus nigrofemoratus*

56.b. Face broader than eye width; scutellum without bristles. ❯ *Platycheirus lundbecki*

Note: *Platycheirus brunnifrons* and *Platycheirus goeldlini*: females unknown, but may key out here.

57.a. Thorax: posterior anepisternum entirely dusted; sternite 2: hairs long, at least in front; frons with small but obvious dust spots, but sometimes reduced and only present along eye margin; larger species: 7-9 mm ❯ 58

57.b. Thorax: lower part of posterior anepisternum shining; sternite 2: hairs very short; frons without dust spots, at most eye margin dusted;; abdomen: figure 651; smaller species: 6-7 mm ❯ *Platycheirus sticticus* Meigen

58.a. Tergite 3: about 2 times wider than long; tergites 3 and 4 with well marked grey spots; abdomen: figure 650 ❯ *Platycheirus albimanus* Fabricius

58.b. Tergite 3: about as long as wide; tergites 3 and 4 with small, vaguely marked spots ❯ *Platycheirus laskai* Nielsen

figure 650. *Platycheirus albimanus*, abdomen of female (Verlinden).

figure 651. *Platycheirus sticticus*, abdomen of female (Verlinden).

figure 652. *Platycheirus manicatus*, abdomen of female (Verlinden).

59.a. Antennae: 3rd segment black ❯60
59.b. Antennae: 3rd segment partly yel-lowish-orange, at least below ❯ 74

figure 653. *Platycheirus tarsalis*, abdomen of female (Verlinden).

60.a. Face, in particular the mouth rim, strongly projected forwards (figure 610); hind femora black at base ❯ 61
60.b. Face flattened in lateral view, the facial knob and the mouth rim project-ing forwards equally; hind femora for most species yellow at base, for a few species black (only for *P. angustipes, P. podagratus, P. ramsarensis*) ❯ 64

61.a. Thoracic dorsum shiny; frons not completely dusted to entirely shiny, at least shiny on either side of ocellular tri-angle and on post-ocular strip ❯ 62
61.b. Thoracic dorsum dull; frons entirely dusted; abdomen: figure 652 ❯ *Platycheirus manicatus* Meigen

62.a. Frons shiny, at most dusted along eye margin ❯ 63
62.b. Frons extensively dusted; abdomen: fig-ure 653 ❯ *Platycheirus tarsalis* Schummel

63.a. Frons dusted along eye margin ❯ *Platycheirus melanopsis* Loew
63.b. Frons undusted, entirely shiny ❯ *Platycheirus tatricus* Dusek and Laska

64.a. Femur 3 black at base, leg 3 black with yellow knee [*P. ramsarensis*: femur 1 with the hairs on the postero-lateral sur-face confined to the basal 2/3 or less of the length of the femur] ❯ 65
64.b. Femur 3 pale at base (check bases carefully), tibia 3 yellow or yellow with black ring; tergites 3 and 4 with moder-ately large to large spots; femur 1 with the hairs on the postero-lateral surface continuing to the tip of the femur ❯ 67

65.a. Femur 1 with the hairs on the postero-lateral surface continuing to the tip
of the femur **>** 66

65.b. Femur 1 with the hairs on the postero-lateral surface confined to the basal
2/3 or less of the length of the femur;
pale markings on the tergites may be
brownish, ill-defined and obscured by
dusting; dusting on post-ocular strip
interrupted at either side of the ocellar
triangle; some of segments of tarsus 1
and 2 distinctly brown, contrastingly
darker than the tibiae, which are more
yellow **>** *Platycheirus ramsarensis*
Goeldlin, Maibach and Speight

66.a. Antero-lateral surface (at least) of
femur 1 entirely pale, yellowish; pale
marks (when present) on tergites 3 and
4 lengthening noticeably towards the
mid-line, they are small round or triangular spots, well-removed from side
margin **>** *Platycheirus podagratus*
Zetterstedt

Jizz: at first sight similar to *Melanostoma. P. europeus*
can be very black on femur 3, but it has larger, rectangular spots approaching the side margin.

66.b. Femur 1 black/dark brown on its
entire circumference for most of its
length; tergites 3 and 4: spots rectangular or rounded **>** *Platycheirus angustipes*
Goeldlin

67.a. Tergite 6 black or at most with very
small yellow spots; femur 3 orange with
black ring at mid-length or almost completely black**>** 68

67.b. Tergite 6 with extensive yellow
markings or completely yellow; femur 3
either completely yellow or with a black
ring at mid-length **>** 71

68.a. Abdomen more than 2.7 times as
long as wide; sternite 2 longer than
wide; frons: dust restricted to 2 well separated triangular spots (figure 654);
wing: cell bm usually with a bare spot at
base **>** 69

68.b. Abdomen at most 2.7 times as long
as wide; sternite 2 wider than long;
frons: in addition to the triangular spots,
a dust spot directly above the antennae,
all dust spots may be merged (figure
655, figure 656); wing completely trichose **>** 70

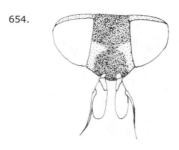
654.

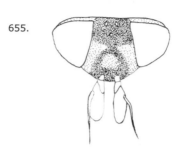
655.

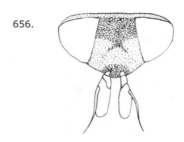
656.

figure 654. *Platycheirus europeus*, head of
female, top view.

figure 655. *Platycheirus occultus*, head of
female, top view.

figure 656. *Platycheirus clypeatus*, head of
female, top view (Van der Linden).

69.a. Notopleurae and anepisternum
dusted; spots on tergites 2- 4 yellowish;
tergite 5 with a triangular pair of spots
or only with dust markings; abdomen
not strongly narrowed to the tip (figure
657) **>** *Platycheirus europaeus* Goeldlin,
Maibach and Speight

69.b. Notopleurae and anepisternum
shiny; spots on tergites 2-4 reddish; tergite 5 without spots; abdomen strongly
narrowed to the tip (figure 658) **>**
Platycheirus angustatus Zetterstedt
Jizz: abdomen slender and pointed.

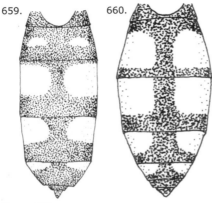

657.   658.

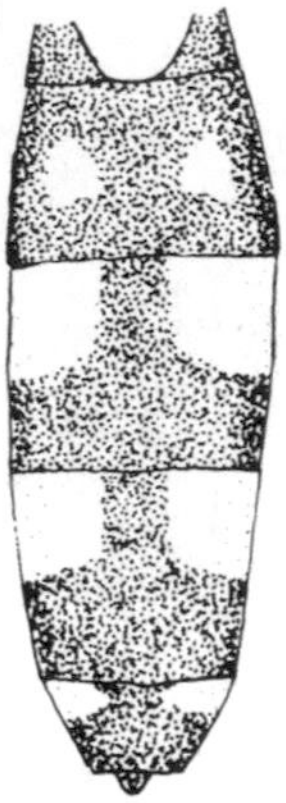

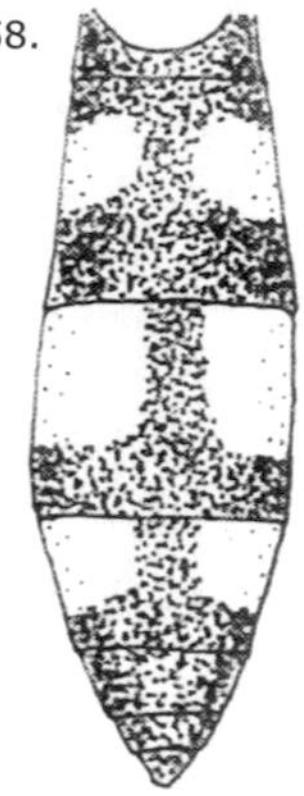

figure 657. *Platycheirus europeus*, abdomen
of female.
figure 658. *Platycheirus angustatus*,
abdomen of female (Van der Linden).

659.   660.

figure 659.
Platycheirus occultus, abdomen of female.
figure 660. *Platycheirus clypeatus*, abdomen
of female (Van der Linden).

70.a. Tergites 3 and 4: spots separated
from the front margin of tergite, rounded, about 1/2 the length of the tergite
(figure 659); frons: no dusting behind
the ocellar triangle, dust spots well separated from each other (figure 655) **>**
Platycheirus occultus Goeldlin, Maibach
and Speight
Jizz: spots on tergite 3 (-4) separated from front
margin.
70.b. Tergites 3 and 4: spots reach the
front margin of the tergite, rectangular,
about 2/3 the length of the tergite (figure 660); frons: dusting of eye rim con-
tinues behind the ocellular triangle, dust
spots on frons usually merged (figure
656) **>** ***Platycheirus clypeatus*** Meigen

71.a. Femur 3 yellow to orange, without
extensive black markings **>** 72
71.b. Femur 3 with dark ring at mid-
length **>** 73

72.a. Frons: dust spot above the antennae
absent or small, if present then well separated from the triangular dust spots
near the eyes; abdomen: figure 661;
antennae: 3rd segment large and square
(figure 662) **>** ***Platycheirus fulviventris***
Macquart
72.b. Frons: dust spot above antennae
merged with the triangular dust spots
near the eye; antennae: 3rd segment
smaller and rounded (figure 663) **>**
Platycheirus perpallidus Verrall

73.a. Femur 1 with a long white hair subbasally; posterior anepisternum dusted;
tergite 2 with smaller yellow spots (figure 664) **>** ***Platycheirus immarginatus***
Zetterstedt
73.b. Femur 1 without a white hair subbasally; posterior anepisternum shiny on
lower 1/2, dusted on upper 1/2; tergite
2 with larger yellow spots (figure 665) **>**
Platycheirus scambus Staeger

figure 661. *Platycheirus fulviventris*,
abdomen of female (Verlinden).

figure 662. *Platycheirus fulviventris*, antenna
of female (Verlinden).

figure 663. *Platycheirus perpallidus*, antenna of female (Verlinden).

74.a. Face flat, facial knob and mouth rim equally protruding **❭** 75
74.b. Face protruding, mouth rim strongly projected forwards **❭** *Platycheirus complicatus* Becker

75.a. Notopleural lobes shiny; abdomen with 3 (seldom 4) pairs of spots, spots on tergite 2 like a comma; abdomen 1.9–2.3 times as long as wide; abdomen with elongate spots (figure 666) **❭** 76
75.b. Notopleural lobes dusted; abdomen with 4 pairs of spots; abdomen 1.6–1.7 times as long as wide **❭** 77

76.a. Face: colour of dust above and below antennal implant is the same, dustless median facial stripe not broader than smallish facial knob; face with scarcely developed mouth rim **❭** *Platycheirus scutatus* Neigen

figure 666. *Platycheirus scutatus*, abdomen of female (Verlinden).

76.b. Face: colour of dust above antennal implant much darker than dust below it; dustless median facial stripe slightly broader than strong facial knob; face with strongly developed mouth rim **❭** *Platycheirus splendidus* Rotheray
Note: females of these species cannot be reliably separated, female *P. aurolateralis* Stubbs unknown, but presumably very similar to *P. splendidus*.

77.a. Anepimeron with long hairs, evenly spread, wavy on top, and never forming a compact tuft. Longest katepisternal hairs at least 1/2 as long as arista. Pleurae shiny, faintly dusted **❭** 78
77.b. Anepimeron with long hairs confined to dorsal part, sometimes forming a compact tuft, and sometimes with a few shorter hairs on ventral part. Longest katepisternal hairs short, about 1/4 as long as arista. Pleurae dull, densely dusted **❭** 79

78.a. Femora 1–3 with black ring. 1st and 2nd antennal segments black. Frons with well-defined triangular pollinose spots, occupying less than 1/3 of frons. Facial knob projecting, well-defined. Anepimeron completely covered with long hairs. Hairs on anepisternum reach backwards as far as the end of the anepimeron. Abdominal spots all well separated from anterior margins of the tergites (figure 667) **❭** *Platycheirus parmatus* Rondani

664.   665.

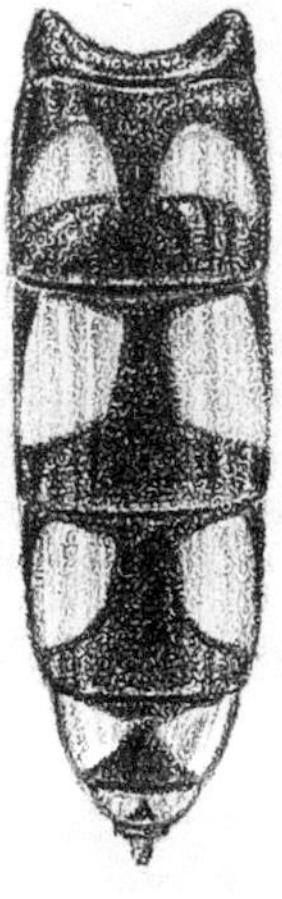

figure 664. *Platycheirus immarginatus*, abdomen of female.
figure 665. *Platycheirus scambus*, abdomen of female (Verlinden).

figure 667. *Platycheirus parmatus*, abdomen of female (Verlinden).

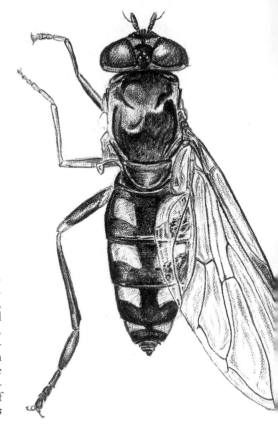

figure 668. *Platycheirus peltatus*, habitus of female (Verlinden).

78.b. Femora 1 and 2 without black rings. 1st and 2nd antennal segments orange. Frons with ill-defined pollinose spots, occupying about 1/3 of frons. Facial knob inconspicuous and ill-defined. Anepimeron without long hairs on ventral 1/4-1/3. Hairs on anepisternum reach backwards to the middle of the anepimeron. Abdominal spots on tergites 3 and 4 reach anterior margins of the tergites ❯ *Platycheirus jaerensis* Nielsen

79.a. Base of all femora black. Hairs on thoracic dorsum and scutellum black, except in front of transverse suture where they are silvery. Spots on tergites 2-5 strongly reduced, ill-defined, brownish. Exterior fringe of femur 1 with long black hairs ❯ *Platycheirus islandicus* Ringdahl
79.b. Femora 1 and 2 yellow, as well as base of femur 3. Hairs on thoracic dorsum and scutellum predominantly or entirely yellow. Spots on abdomen, at least on tergites 3-5, well-defined, yellow to reddish. Exterior fringe of femur 1 with medium-sized yellow hairs ❯ 80

80.a. Spots on tergites 3 and 4 subrectangular, sometimes separated from anterior margins of the tergites, anterior (and posterior) margins of the spots almost straight. Spots on tergite 2 kidney-shaped to semicircular, well separated from anterior margins of the tergites, sometimes reduced and ill-defined ❯ 81

80.b. Spots on tergites 3 and 4 trapezium-shaped, anterior margins of spots convex, separated from anterior margins of the tergites in lateral corners of the spots, posterior margins of the spots concave, sometimes reduced. Spots on tergite 2 subtriangular close to anterior margin of the tergite, sometimes reduced and ill-defined ❯ 82

81.a. Anepimeron with the long hairs on dorsal part forming a compact tuft with a silvery shine, hair base not visible, ventral part bare. Frons with clearly defined triangular greyish pollinose spots, occupying about 1/3 of frons. Spots on tergites 3 and 4 about twice as broad as long. 3rd tergite about twice as broad as long ❯ *Platycheirus holarcticus* Vockeroth

81.b. Anepimeron with the long hairs on dorsal part less dense, hair bases visible, ventral part with a few scattered hairs. Frons with broad and ill-defined pollinose spots, occupying more than 3/4 of frons. Spots on tergites 3 and 4 about 1.5 times as broad as long. 3rd tergite distinctly less than twice as broad as long **> *Platycheirus amplus*** Curran

82.a. Tibia 2 uniformly broadened from base towards apex then becoming smaller and swelling strongly on apical 1/5 **> *Platycheirus nielseni*** Vockeroth
82.b. Tibia 2 uniformly broadened from base to apex, sometimes with a discrete swelling on apical 1/5 **> 83**

83.a. Tibia 2 with a discrete swelling on apical 1/5. Dusting on frons more reduced and well-defined. Extremity of the abdomen pointed due to the relatively narrow posterior margin of tergite 5 and the proportionally long and narrow tergites 6-8 **> *Platycheirus nielseni*** Vockeroth
83.b. Tibia 2 uniformly broadened from base to apex, sometimes with a modest swelling on apical 1/5. Dusting on frons more extended and not well-defined. Extremity of the abdomen blunt, due to the relatively broad posterior margin of tergite 5 and proportionally short and broad tergites 6 and 7 (figure 668) **> *Platycheirus peltatus*** Meigen

POCOTA

Introduction

The sole *Pocota* species is a convincing mimic of the bumblebees *Bombus terrestris* and *Bombus lucorum*. It is a rare woodland species. Larvae live in moist, decaying wood.

Key

1. Single European species. Large bumblebee mimic: front part of thorax with yellow hairs, hind part and scutellum with black hairs, front part of abdomen with black hairs, tergite 3 reddish-yellow hairs and hind tergites with whitish-yellow hairs; wing: veins in the middle darkly infuscate. 12-13 mm. Central Europe **> *Pocota personata*** Harris
Jizz: head is relatively small, smaller than thorax.

PSARUS

Introduction

Psarus hoverflies are almost hairless with a flattened, red abdomen and long antennae. They occur in well-drained old growth *Quercus* forest with a diverse herb layer. They fly along woodland paths and forest edges. Males have been seen sitting in the sun on the ends of dead branches of trees growing beside paths, 2 metres or more above ground level, and repeatedly returning to particular branches (Speight, 2003).

figure 669. *Psarus abdominalis*, habitus of male (Verlinden).

Key

1. Single European species. Thoracic dorsum and scutellum black with coarse punctuation, almost dull; legs: femora black, tibiae and tarsae brown; abdomen largely red, black on base and tip (figure 669). 8-10 mm. Central Europe > ***Psarus abdominalis*** Fabricius

Jizz: red abdomen and long antennae unique.

PSILOTA

Introduction

Psilota occurs in a range of forests, from thermophilous *Quercus pubescens* to damp *Fagus* forest and *Fagus/Picea* forest, up into the *Picea* zone. Adults are largely arboreal and may be found in small glades within forests (Speight, 2003). They have been seen visiting willow catkins, hawthorn and apple blossoms. Females can be found by searching shady tree trunks. Their larvae have been found in decaying sap under bark (Speight, 2003).

Recognition

Small, black hoverflies with long wings and a rounded abdomen, without venia spuria in the wing. *Psilota* do not have a facial knob and only the mouth edge protrudes. When resting, they appear much like other small forest Diptera. In flight, they behave more like a typical hoverfly. There are three species in our region, including *Psilota atra* Fallén. The key in the previous edition was based on the prepublication of Smit and Zeegers (used with the consent of Smit and Zeegers). Because Smit and Zeegers (2005) switched *P. atra* and *P. anthracina* in their final publication, the key in the previous edition switched both species. The present key follows the final publication of Smit and Zeegers (2005).

Key

1.a. Thoracic dorsum with black hairs; face black shiny; wing entirely hyaline, longitudinal veins at the base yellow, much paler than distal parts > 2
1.b. Thoracic dorsum with white hairs; face with white dusting; wing slightly infuscated centrally, longitudinal veins at the base slightly paler than more distal parts. 5-6 mm. Central and Southern Europe > ***Psilota inupta*** Rondani

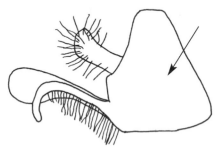

figure 670. *Psilota anthracina*. Male genitalia (J.T. Smit).

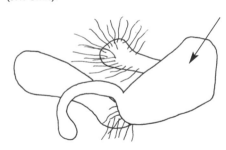

figure 671. *Psilota atra*. Male genitalia (J.T. Smit).

figure 672. *Psilota atra*, habitus of male (Verlinden)

2.a. Femur 3 only slightly thickened, about as broad as femur 2. Female: posterior anepisternum black-haired; abdomen: 3rd tergite predominantly black-haired; 3rd antennal segment about 1.5 times longer than wide. Male: genitalia: epandrium in lateral view distinctly higher than long (figure 670). 5-6 mm. Confused with next species, probably Northern and Central Europe **> Psilota anthacina** Meigen

2.b. Femur 3 strongly thickened, twice as broad as femur 2. Female: posterior anepisternum light-haired; abdomen: predominantly light haired, the 2nd tergite with conspicuous long white hairs in the frontal corners, 3rd tergite predominantly light-haired; 3rd antennal segment about 2 times longer than wide. Male: genitalia: epandrium in lateral view distinctly longer than high (figure 671). 5-6 mm. Confused with previous species, probably Central Europe, Spain, Greece **> Psilota atra** Fallén

PYROPHAENA

Introduction

Pyrophaena occur in habitats with open water nearby. *P. granditarsa* prefers open standing water, such as ditches, where it flies around the vegetation. *P. rosarum* occurs in marshy environments, such as bogs and marshes fed by groundwater. The larvae are carnivorous. *Pyrophaena* is now placed *Platycheirus*.

Key

1.a. Tergites 2-4 largely reddish, in females the hind corners of tergites 2 and 3 black (figure 673, figure 674). Male: metatars 1 with a protuberance. 8-11 mm. Europe, in Asia to Pacific coast, Nearctic **> Pyrophaena granditarsa** Forster
Jizz: slender species, with blackish wings and reddish abdomen.

1.b. Tergite 2 black, tergites 3 and often 4 with a whitish to yellowish band (figure 675), pale specimens may develop 2 pale spots on tergite 2. Male: metatars 1 without protuberance. 7-10 mm. Europe, Siberia, Nearctic **> Pyrophaena rosarum** Fabricius
Jizz: slender, blackish species with violet sheen on the wings.

figure 673. *Pyrophaena granditarsa*, habitus of female (Verlinden).

figure 674. *Pyrophaena granditarsa*, abdomen of male (Verlinden).

figure 675. *Pyrophaena rosarum*, habitus of male (Verlinden).

RHINGIA

Introduction

Rhingia campestris larvae live in cow dung and the adult flies are often numerous in and near agricultural areas. *R. campestris* adults forage on flowers with deep nectaries, such as Lamiaceae. The larvae of the other species probably also live in the dung of large herbivores (Van Steenis, 1998). *Rhingia borealis* appear to prefer elevated areas (from hills to mountains) where they occur in open areas in scattered woodlands with tall herbaceous vegetation. They are often encountered on leaves.

Recognition

The long snout and reddish abdomen make this genus easy to recognise in the field. The key follows Van Steenis (1998a).

Key

1.a. Thoracic dorsum with grey dusting; scutellum light brownish, translucent; arista bare or haired **>** 2

1.b. Thoracic dorsum and scutellum black; thoracic dorsum densely covered with long hairs; arista with long hairs; snout short and curved downwards (as figure 676); tergites with black hind and side margins. 6-8 mm. Northern and Central Europe, in Asia into Siberia **>** male *Rhingia borealis* Ringdahl (= *Rhingia austriaca* auctorum)
Jizz: small and pale *Rhingia* found in woodlands.

2.a. Snout shorter and curved downwards, as long as or shorter than the horizontal diameter of an eye; tergites only with dark hind margin; arista either bare or long-haired **>** 3

2.b. Snout long and straight, longer than the horizontal diameter of an eye (figure 677); tergites with black hind and side margins, often with a black marking in the middle of the tergite (figure 678); arista with short hairs; tibia red. 8-11 mm. Europe, in Asia east to Pacific coast **>** *Rhingia campestris* Meigen
Jizz: large, with round, reddish abdomen and long snout.

3.a. Arista bare; antennae: 3rd segment as long as wide; snout hardly curved, as long as the horizontal diameter of an eye (figure 679); thoracic dorsum mostly dusted; tibia red. 8-9 mm. Central Europe, in Asia into Siberia **>** *Rhingia rostrata* Linnaeus.

3.b. Arista with long hairs; antennae: 3rd segment 1.5 times as long as wide; snout short and strongly curved, shorter than the horizontal diameter of an eye (figure 676); thoracic dorsum shiny, with dusted longitudinal stripes; tibia 3 black at apex. 6-8 mm. Northern and Central Europe, in Asia into Siberia. Males in 1.b. **>** female *Rhingia borealis* Ringdahl (= *Rhingia austriaca* auctorum)
Jizz: small and pale *Rhingia* found in woodlands.

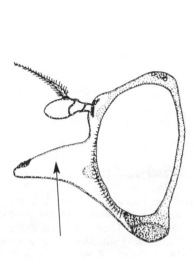

figure 676. *Rhingia borealis*, head of female
(Van Steenis, 1998).

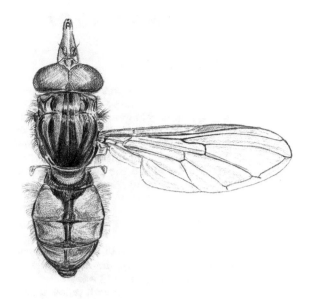

figure 678. *Rhingia campestris*, habitus of
male (Verlinden).

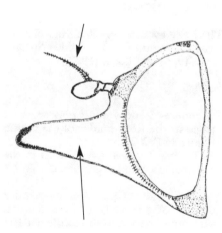

figure 677. *Rhingia campestris*, head of
female (Van Steenis, 1998).

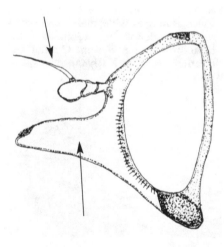

figure 679. *Rhingia rostrata*, head of female
(Van Steenis, 1998).

RIPONNENSIA

Introduction

Riponnesia are found in a variety of damp habitats, such as damp forests, near seepages and springs in grasslands or near running water. Adults visit the flowers of Apiaceae, Ranunculaceae and Rosaceae. The larvae are semiaquatic. Maibach et al. (1994a) established the genus and give a description of most species.

Key

1.a. Legs entirely metallic green; thoracic dorsum with 2 longitudinal stripes of white dust in front 1/2 ❯ 2
1.b. Legs golden green with yellow knees; thoracic dorsum without stripes of white dust. Known from a single male ❯ *Riponnensia insignis* Loew
Jizz: golden green *Riponnensia*.

2.a. Antennae: 3rd segment oval, at most twice as long as wide; male: cerci shorter and broader, without tooth above the basal broadening. 8 mm. Central and Southern Europe ❯ *Riponnensia splendens* Meigen

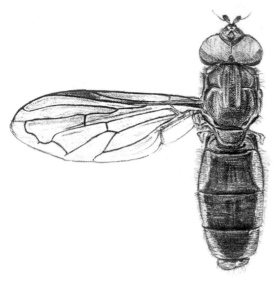

figure 680. *Riponnensia splendens*, habitus of male (Verlinden).

2.b. Antennae: 3rd segment slender, 3 times as long as wide; male: cerci long and slender, with a tooth just above the basal broadening. 8 mm. Southern Europe ❯ *Riponnensia longicornis* Loew

SCAEVA

Introduction

Scaeva are robust hoverflies, with an oval abdomen that is black with white or yellow markings. They are good flyers and travel for long distances. The Southern European *Scaeva dignota* is occasionally encountered as far north as The Netherlands. For this reason the southern *Scaeva mecogramma* is included in the key. *Scaeva pyrastri* may be a migratory species. Their larvae are aphid predators.

Recognition

The hairy eyes and swollen frons characterise the genus. The key follows Speight et al. (1986) and Lucas (1992).

Key

1.a. Tergites 3 and 4: spots oblique, at an angle to the front border of the tergite (figure 681) ❯ 2
1.b. Tergites 3 and 4: spots parallel to the front margin of the tergite (figure 682) ❯ 3

2.a. Front border of spots concave, spots like commas (figure 684); thoracic dorsum: sides only somewhat yellowish; femur 1: at least some black bristles present among the yellow ones. 10-15 mm. Europe, North Africa, in Asia east to Pacific coast, Nearctic ❯ *Scaeva pyrastri* Linnaeus
Jizz: large and shiny, spots white and oblique.
2.b. Front border of spots straight, spots like oblique bars; thoracic dorsum: sides bright yellow; femur 1: only yellow bristles present. 10-15 mm. Southern Europe, North Africa, in Asia into Mongolia, migrates to Central and Northern Europe ❯ *Scaeva albomaculata* Macquart
Jizz: large and shiny, spots yellow and straight, obliquely placed.

681.

682.

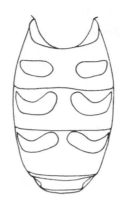

figure 681. *Scaeva pyrastri*, part of abdomen.
figure 682. *Scaeva dignota*, abdomen
(Speight et al., 1987).

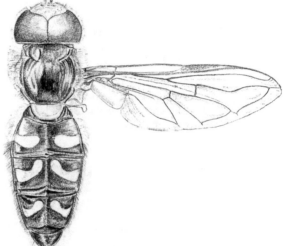

3.a. Males ❯ 4
3.b. Females ❯ 6

4.a. Eye: facets on the anterodorsal side of
the eye distinctly larger than on other
parts; tergites 3 and 4: spots always separated ❯ 5
4.b. Eye: size of all facets equal; tergites 3
and 4: spots generally confluent, forming a band. 10-13 mm. Southern
Europe ❯ *Scaeva mecogramma* Bigot

5.a. Frons strongly inflated, angle between
eyes on frons about 130° (figure 685);
spots do not reach the lateral margin of
the tergites (figure 687). 12-15 mm.
Europe, North Africa, in Asia to Pacific
coast ❯ *Scaeva selenitica* (Meigen)
Jizz: large and shiny, spots yolk yellow.
5.b. Frons weakly inflated, angle between
eyes on frons about 90° (figure 686);
spots normally reach the lateral margin
of the tergites. 10-14 mm. Central and
Southern Europe, North Africa,
migrants disperse further north ❯ *Scaeva
dignota* Rondani
Jizz: smaller than *selenitica*, spots pale
yellow.

6.a. Tergites 3 and 4 with pairs of spots ❯ 7
6.b. Tergite 3 (and normally 4) with a
band (figure 684). 10-13 mm. Southern
Europe ❯ *Scaeva mecogramma* (Bigot)

figure 683. *Scaeva pyrastri*, habitus of male
(Verlinden).

figure 684. *Scaeva mecogramma*, abdomen
(Speight et al., 1987).

685.

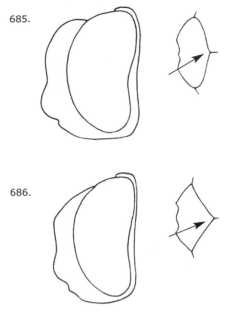

686.

figure 685. *Scaeva selenitica*, face of male.
figure 686. *Scaeva dignota*, face of male
(Speight et al., 1987).

7.a. Frons convex (figure 688); frontal
bristles at the bases of the antennae
black and numerous; spots do not reach
the lateral margin of the tergites (figure
687). 12-15 mm. Europe, North Africa,
in Asia to Pacific coast ❭ *Scaeva selenitica* Meigen

7.b. Frons concave (figure 689); frontal
bristles at the bases of the antennae
white; spots normally reaching the lateral margin of the tergites. 10-14 mm.
Central and Southern Europe, North
Africa, migrants disperse further north ❭
Scaeva dignota Rondani

688.

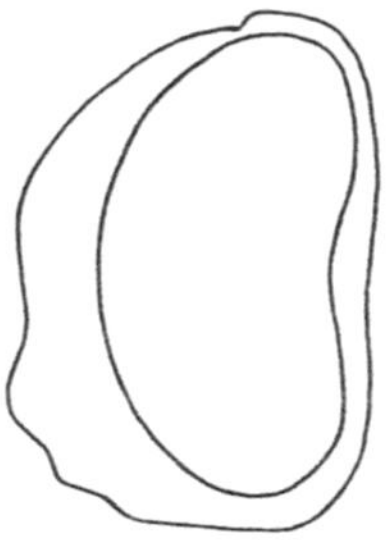

689.

figure 688. *Scaeva selenitica*, face of female
(Speight et al., 1987).
figure 689. *Scaeva dignota*, face of female
(Speight et al., 1987).

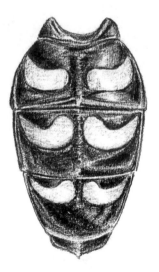

figure 687. *Scaeva selenitica*, abdomen
(Verlinden).

SERICOMYIA

Introduction

Broad-bodied Syrphids that are black with yellow markings. They occur in wetlands and bogs. Face extended downwards, antennae with long-haired arista. The larvae have a long tail for respiration and live in small ponds, rich in decomposing vegetation. The key is based on Van der Goot (1981) and Nielsen (1997).

Key

1.a. Tip of abdomen black (figure 690) > 2

1.b. Tip of abdomen yellow (figure 691, male: pregenital segment; female: tergites 5 and 6); legs yellowish, base of femora black. 14-15 mm. Northern and Central Europe, in Asia into Japan > *Sericomyia silentis* Harris

Jizz: very large, spots and legs yellow.

2.a. Halteres: knob pale > 3

2.b. Halteres: knob black; legs reddish, femur darkened at base; tergites 3 and 4: whitish-yellow bands small, less than 1/3 tergite length, and with a short interruption in the middle (figure 690); scutellum reddish. 12-14 mm. Northern and Central Europe, in Asia to Pacific coast > *Sericomyia lappona* Linnaeus

Jizz: smaller, legs more reddish and spots paler compared to *S. silentis*.

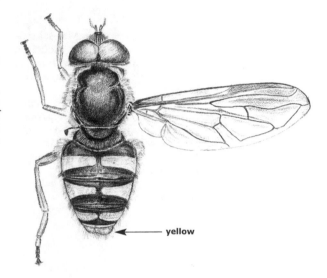

figure 691. *Sericomyia silentis*, habitus of male (Verlinden).

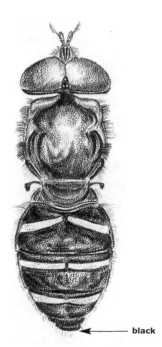

figure 690. *Sericomyia lappona*, habitus of male (Verlinden).

figure 692. *Sericomyia arctica*, abdomen (after Van der Goot, 1981).

3.a. Scutellum dark reddish-brown, different in colour to thoracic dorsum; tergites 3 and 4: pale bands broadly interrupted in the middle, never touching. Female: tergite 5 black **>** 4

3.b. Scutellum black, same colour as thoracic dorsum; tergites 3 and 4: pale bands just interrupted in the middle, bands normally touching. Female: tergite 5 with pale spots. 15-16 mm. Northern Europe, Siberia **>** *Sericomyia nigra* Portchinsky
Jizz: black scutellum and virtually uninterrupted pale bands unique.

4.a. Tibia and tars 3: reddish. Males: genitalia, ventral view: tip of hypandrium rounded. 13-14 mm. Northern Europe, Siberia, Nearctic **>** *Sericomyia arctica* (Schimmer)

4.b. Tibia and tars 3: mainly black. Males: genitalia, ventral view: tip of hypandrium pointed and hooked. 13-14 mm. Eastern Siberia, northern parts of Norway and Sweden **>** *Sericomyia jakutica* Stackelberg
Note: Nielsen (1997) remarks that the colour of tibia and tars 3 is variable.

SPHAEROPHORIA

Introduction

Sphaerophoria hoverflies are small to large, elongated flies. Their bodies are black with yellow markings. The common *S. scripta* can be found everywhere. Other species confine themselves more to heathlands (e.g. *S. virgata*), bogs (e.g. *S. potentillae*) or pine forest (e.g. *S. batava*). Their larvae prey on aphids. *S. scripta* females can be found ovipositing near aphid colonies on a variety of weedy plants.

Recognition

The genus contains a small number of species that are relatively easy to identify,

but the remaining species are difficult. In the present key, the first group consists of *S. estebani, S. loewi, S. rueppellii, S. shirchan* and *S. scripta*. The first four have a yellow stripe only at the front half of the thorax, the abdomen of the latter projects beyond the wings.

All other species can only be identified by the genital characteristics of the males, making it necessary to prepare the genitalia when pinning a specimen. A simple upwards turn of the genitalia suffices. A reference collection can be very useful for gaining confidence with these identifications.

The key is based on a study of collection specimens and Van der Goot (1986), Goeldlin de Tiefenau (1989), Goeldlin de Tiefenau (1991), Verlinden (1991), Claussen (1984), Stubbs (1996), Torp (1994) and Bartsch et al (2009a).

Key

1.a. Thorax with yellow side stripe interrupted above wing base; postalar lobes black, sometimes only darkened (figure 693) **>** 2

1.b. Thorax with continuous side stripe; postalar lobes yellow (figure 694) **>** 5

2.a. Antennae yellow **>** 3

2.b. Antennae black; habitus figure 695, figure 696. 7-9 mm. Northern and Central Europe, in Asia into Mongolia **>** *Sphaerophoria loewi* Zetterstedt
Jizz: dark antennae visible from a distance.

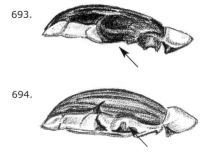

693.

694.

figure 693. *Sphaerophoria rueppelli*, side of thoracic dorsum.
figure 694. *Sphaerophoria scripta*, side of thoracic dorsum (Verlinden).

695.   696.

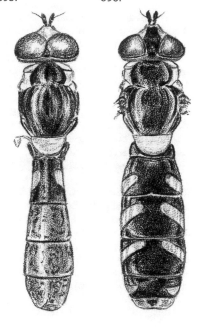

697.

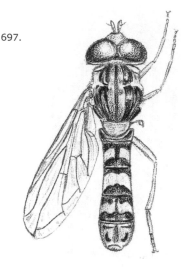

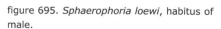

figure 695. *Sphaerophoria loewi*, habitus of male.

figure 696. *Sphaerophoria loewi*, habitus of female (Verlinden).

698.

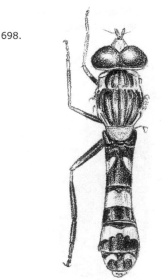

3.a. Face yellow below antennae, at most facial knob vaguely blackened **>** 4

3.b. Face with distinct black median stripe, running from the antennae down to the mouth rim. Central Europe, Siberia **>** ***Sphaerophoria shirchan*** Violovitsh

4.a. Abdomen constricted at segments 2 and 3 (take care with dried specimens with curved margins!), smaller than segments 5 and 6; tergites 3-5 normally with yellow spots, which narrow towards the tergite margin and may merge to form a broad band (figure 697, figure 698). 5-8 mm. Central and Southern Europe, North Africa, Ethiopian region, in Asia to Pacific coast **>** ***Sphaerophoria rueppellii*** (Wiedemann)

Jizz: abdomen constricted at the base, tergites 4-6 usually with much orange-yellow patterning in addition to the yellow spots.

figure 697. *Sphaerophoria rueppelli*, habitus of male.

figure 698. *Sphaerophoria rueppelli*, habitus of male (Verlinden).

4.b. Males only. Abdomen elongate, with parallel margins; tergites 2-5 normally with elongate, yellow bands of constant width, which are sometimes interrupted in the middle. 5-8 mm. Mountains of Central Europe **>** ***Sphaerophoria estebani*** Goeldlin de Tiefenau

Jizz: linear lemon yellow bands on shiny black tergites.

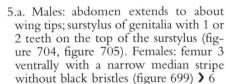

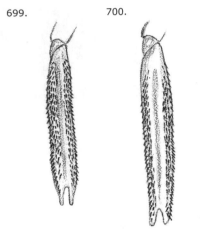

figure 699. *Sphaerophoria interrupta*, femur 3 of female, ventral view.
figure 700. *Spaerophoria scripta*, femur 3 of female, ventral view (Verlinden).

5.a. Males: abdomen extends to about wing tips; surstylus of genitalia with 1 or 2 teeth on the top of the surstylus (figure 704, figure 705). Females: femur 3 ventrally with a narrow median stripe without black bristles (figure 699) **❯** 6

5.b. Males: abdomen extends well beyond wing tips (figure 701); surstylus of genitalia without teeth (figure 703). Females: femur 3 ventrally with a broad median stripe without black bristles (figure 700). 9-12 mm. Europe, North Africa, in Asia to Pacific coast, highly migratory **❯** ***Sphaerophoria scripta*** Linnaeus
Jizz: abdomen longer than closed wings.

6.a Surstylus in lateral view with a broad lobe bearing a slender tooth: on both lobes the width of this tooth (measured at the base) 1/4 or less of the width of the lobe (figure 704), a 2nd, small process may be present on the inner side of surstylus; left and right surstyli symmetrical or slightly asymmetrical **❯** 7

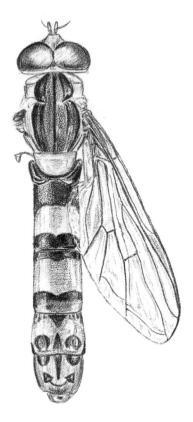

figure 701. *Sphaerophoria scripta*, habitus of male (Verlinden).

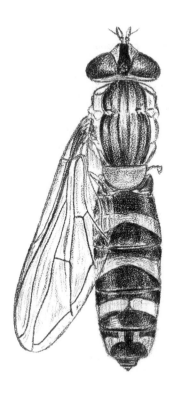

figure 702. *Sphaerophoria scripta*, habitus of female (Verlinden).

figure 703. *Sphaerophoria scripta*, genitalia of male (Verlinden).

toothed lobe

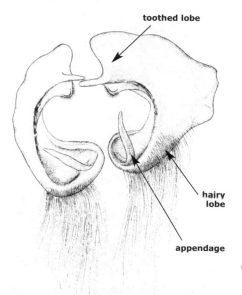

hairy lobe

appendage

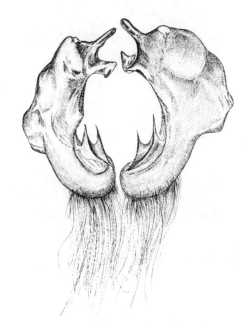

figure 706. *Sphaerophoria philantha*, genitalia of male (Verlinden).

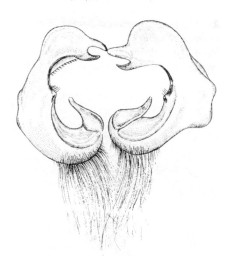

figure 704. *Sphaerophoria virgata*, genitalia of male.

figure 705. *Sphaerophoria fatarum*, genitalia of male (Verlinden).

6.b. Males only. Surstylus in lateral view with a narrower lobe ending in 1 or 2 teeth: the width of the largest tooth (measured at the base) 1/3 or more of the width of the lobe (figure 705, exceptionally 1 of the lobes with a slender tooth, but then the other with a broad tooth); left and right surstyli somewhat (e.g. *S. batava*) to strongly (e.g. *S. interrupta*) asymmetrical ❯ 10

7.a. Surstylus: in lateral view the long tooth removed from the edge of the lobe, often accompanied by a small tooth on the edge (figure 706); hypopygium seen from the back: hairy lobes narrowly triangular, higher than wide (figure 707) [surstylus often, but not necessarily, heavily chitinised and black] ❯ 8

7.b. Surstylus: on the tip with 1 tooth, in lateral view, this tooth on the edge of the lobe (figure 704); hypopygium seen from the back: hairy lobes almost semicircular, as high as wide or only slightly higher (figure 708) [surstylus often weakly chitinised and yellow] ❯ 9

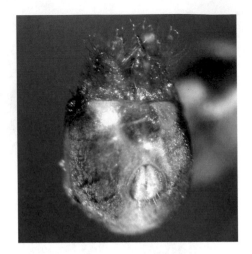

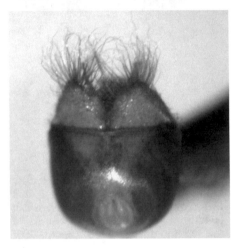

figure 707. *Sphaerophoria philantha,* hypopygium, back view.
figure 708. *Sphaerophoria virgata,* hypopygium, back view (Van Veen).

710.

711.

712.

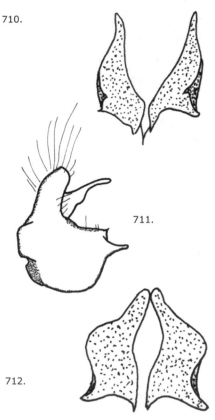

figure 710. *Sphaerophoria philantha*, internal appendage of hairy lobe.
figure 711. *Sphaerophoria boreoalpina*, genitalia of male.
figure 712. *Sphaerophoria boreoalpina*, internal appendage of hairy lobe (after Goeldlin, 1989).

figure 709. *Sphaerophoria philantha*, abdomen of male (Verlinden).

figure 713. *Sphaerophoria virgata*, abdomen of male (Verlinden).

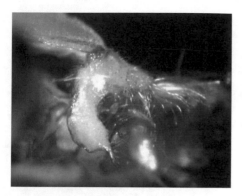

figure 714. *Sphaerophoria bankowskae*, genitalia of male (Van Veen).

715.

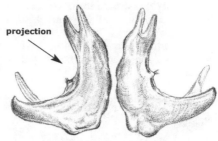

projection

716.

equally large teeth

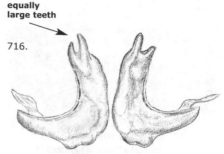

figure 715. *Sphaerophoria taeniata*, genitalia of male.

figure 716. *Sphaerophoria batava*, genitalia of male (Verlinden).

8.a. Long tooth on tip of the surstylus approximately in the middle of the lobe in lateral view (figure 706); toothed lobe in direct comparison with next species smaller, and more angularly bent; appendage on hairy lobe narrowly triangular, gradually tapering (figure 710). 7-8 mm. Northern Europe and northern part of Central Europe ❯ *Sphaerophoria philantha* Meigen
Jizz: rather dark species, often (but not always) pairs of spots on tergites which merge in pale specimens.

8.b. Long tooth on tip of the surstylus located at about 1/4 from the inner edge in lateral view (figure 711); toothed lobe broader, very smoothly bent; inner appendage of hairy lobe broad, suddenly narrowing into tip (figure 712). 7-8 mm. Northern Europe and mountains of Central Europe ❯ *Sphaerophoria boreoalpina* Goeldlin de Tiefenau
Jizz: Darker species than *S. philantha*: yellow spots small.

9.a. Inner process of surstylus is a small semiglobular projection evenly implanted with many small bristles; hairy lobe small in lateral view (figure 704); surstylus yellow. 8-10 mm. Northern and Central Europe ❯ *Sphaerophoria virgata* Goeldlin de Tiefenau

9.b. Inner process of surstylus a high, slender keel; hairy lobe broad in lateral view (figure 714); surstylus often whitish. 8-10 mm. Northern Europe and mountainous parts of Central Europe ❯ *Sphaerophoria bankowskae* Goeldlin

10.a. Inner curve of the surstylus with a distinct projection, which may be prominent on only 1 of the surstyli (figure 715) ❯ 11

10.b Inner curve of surstylus without projection on both surstyli, at most with clustered hairs that are implanted on small elevations (figure 716) ❯ 15

11.a. Genitalia: base of toothed lobe without bulge and, seen from the side, the form of the toothed lobe rectangular, sometimes weakly S-shaped; toothed lobe about as long as hairy lobe; tergite 2: with yellow band (than the projection on the toothed lobe or just at the contact zone of both lobes) or with 2 yellow spots ❯ 12

11.b. Genitalia: base of toothed lobe with a distinct bulge ventrally at the base, the form of the toothed lobe triangular seen from the side; toothed lobe longer than hairy lobe; projection is present on hairy lobe; tergite 2: always with yellow band. 8-10 mm. Northern Europe and Siberia, above polar circle. ❯ *Sphaerophoria pallidula* Mutin

12.a. Surstylus shouldered, broadened below teeth; lobes of surstylus with 1 long tooth and 1 smaller tooth or process, which may be missing on 1 of the surstyli; projection on the hairy lobe **> 13**

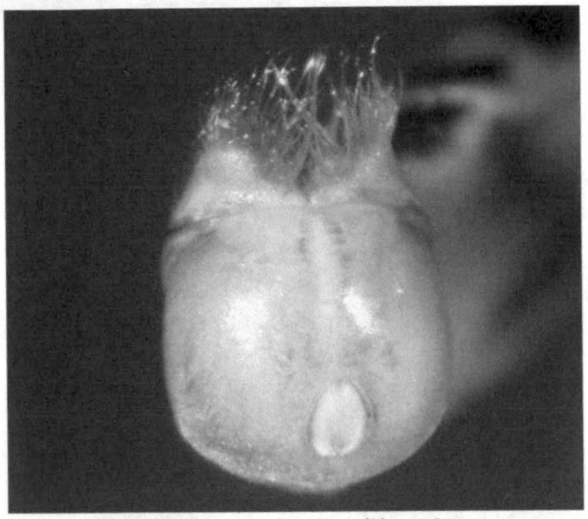

12.b. Surstylus not shouldered, with parallel margins; lobes of surstylus with 2 equally strong teeth, the smallest longer than 1/3 the length of the longest (figure 716); projection at the base of the toothed lobe. 8–10 mm. Central Europe, in Asia to Pacific coast **> Sphaerophoria taeniata** Meigen

Jizz: rather yellow *Sphaerophoria*, on first sight much like *S. scripta*, but wings as long as abdomen.

figure 717. *Sphaerophoria abbreviata*, hypopygium of male, back view (Van Veen).

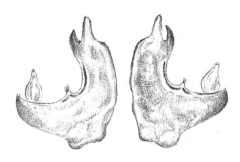

figure 720. *Sphaerophoria fatarum*, genitalia of male (Verlinden).

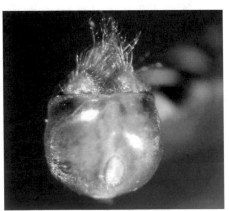

figure 718. *Sphaerophoria abbreviata*, surstylus of male (Van Veen).

figure 721. *Sphaerophoria fatarum*, abdomen of male (Verlinden).

large tooth →

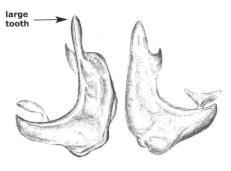

figure 719. *Sphaerophoria fatarum*, hypopygium of male, back view (Van Veen).

figure 722. *Sphaerophoria interrupta*, genitalia of male (Verlinden).

13.a. Hypopygium from below: hairy lobes wider than high or as high as wide; surstyli dark yellow, sometimes with black parts 〉 14
13.b. Hypopygium from below: hairy lobes slightly higher than wide; surstyli whitish 〉 see 9b, *Sphaerophoria bankowskae* Goeldlin

14.a. Appendage of hairy lobe viewed from below or above: not contricted at the base, gradually tapering into the tip; toothed lobe less shouldered, rather slender and its width about twice the width of the tooth at its base (figure 720); Hypopygium from below: hairy lobes wider than high (figure 719); abdomen: figure 721. 7-10 mm. Northern and Central Europe 〉 *Sphaerophoria fatarum* Goeldlin de Tiefenau
Note: In Northern Europe, Siberia, above polar circle: *Sphaerophoria kaa* Violovitsh, which is very similar. The appendage of hairy lobe has a short tapering tip instead of a long one.
14.b. Appendage of hairy lobe viewed form below or above: petiolate with a constriction at the base (like an arrow point); toothed lobe distinctly shouldered, its width more than twice the tooth at its base (figure 718); Hypopygium from below: hairy lobes as high as wide (figure 717). 7-10 mm. Northern and Central Europe 〉 *Sphaerophoria abbreviata* Zetterstedt

15.a. Surstyli both with 2 equally strong teeth, the smallest longer than 1/3 the length of the longest (figure 716) 〉 16
15.b. Surstylus with 1 long, stout tooth, for some species accompanied by 1 smaller tooth or process, the latter shorter than 1/4 the length of the longest and may only be present on 1 of the surstyli (figure 722) 〉 18

16.a. Tergite 2 with a yellow band 〉 17
16.b. Tergite 2 with a pair of spots, at most barely touching 〉 14

17.a. Surstylus: angle between toothed lobe and hairy lobe about 90°; inner curve of surstylus implanted with clustered hairs (figure 716); hypopygium seen from below higher than wide, hairy lobes as high as wide. 8-10 mm. Northern and Central Europe 〉 *Sphaerophoria batava* Goeldlin de Tiefenau

17.b. Surstylus: angle between toothed lobe and hairy lobe wider than 110°; inner curve of surstylus implanted with an evenly spaced row of long white hairs (figure 724); hypopygium seen from below wider than high, hairy lobes wider than high (figure 725). 8-10 mm. Northern Europe and northern part of Central Europe 〉 *Sphaerophoria potentillae* Claussen

18.a. Hypopygium seen from below: each hairy lobe semicircular, not reduced; surstylus: in side view hairy lobe not reduced, a little narrower than the toothed lobe, toothed lobe: large tooth on the inner side of the tip 〉 19
18.b. Hypopygium seen from below: each hairy lobe reduced to a small stripe, which may be hardly visible; surstylus: in side view: hairy lobe narrow almost lacking, toothed lobe: large tooth located on the outside of the tip. 7-10 mm. Northern and Central Europe, in Asia to Pacific coast 〉 *Sphaerophoria chongjini* Bankowska

19.a. Surstylus: appendage of hairy lobe broadly triangular or broadly ovoid; hypopygium seen from below as wide as high or higher than wide 〉 20
19.b. Surstylus: appendage of hairy lobe narrow and linear (figure 728); hypopygium seen from below wider than high, hairy lobes wider than high; hairs on lateral margin of tergites 3-5 black. 7-10 mm. Central Europe 〉 *Sphaerophoria infuscata* Goeldlin de Tiefenau

figure 723. *Sphaerophoria batava*, abdomen of male (Verlinden).

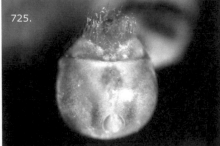

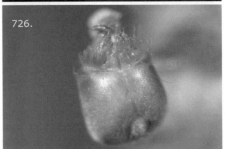

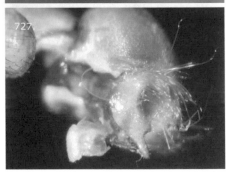

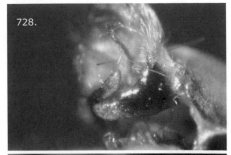

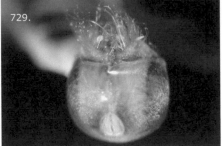

figure 728. *Sphaerophoria infuscata*, surstylus (Van Veen).

figure 729. *Sphaerophoria infuscata*, hypopygium from below (Van Veen).

figure 730. *Sphaerophoria interrupta*, abdomen of male (Verlinden).

figure 724. *Sphaerophoria potentillae*, genitalia of male.

figure 725. *Sphaerophoria potentillae*, hypopygium of male, back view.

figure 726. *Sphaerophoria chongjini*, hypopygium.

figure 727. *Sphaerophoria chongjini*, genitalia of male.

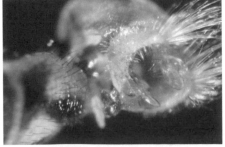

figure 731. *Sphaerophoria laurae*, surstylus of male (Van Veen).

20.a. Surstylus: large tooth accompanied by a small tooth at the base, large tooth somewhat narrowed towards base (figure 722); hairs on lateral margin of tergites 3-5 pale, in melanistic specimens intermixed with a few black; tergite 2-4 with pairs of spots, which may touch in the middle. 7-10 mm. Europe, Siberia **>** ***Sphaerophoria interrupta*** (Fabricius) (= *Sphaerophoria menthastri* auct.)

20.b. Surstylus: small tooth at the base of the large tooth absent, large tooth gradually widened towards base (figure 731); hairs on lateral margin of tergites 3-5 black; tergites 2-4 with yellow bands. 7-10 mm. Northern Europe, mountains of Central Europe **>** ***Sphaerophoria laurae*** Goeldlin de Tiefenau

SPHECOMYIA

Introduction

Sphecomyia are rare, boreal hoverflies that live along rivers and streams in boreal forests. These large, yellow and black hoverflies are excellent mimics of social wasps.

Key

1. Only a single species in Europe. Face yellow with black stripe; tergites black with yellow bands. 12-14 mm. Northern Europe, Northern Siberia **>** ***Sphecomyia vespiformis*** Gorski

Jizz: remains a convincing wasp mimic even when pinned!

SPHEGINA

Introduction

Sphegina inhabit shaded areas in damp coniferous and deciduous forests, in association with running water. Adults fly through the vegetation and are found visiting flowers of white umbellifers and other flowers along the forest edge. The larvae have been found in wet, sappy material beneath a patch of wet bark on a living *Ulmus* (Hartley, 1961). Larvae have also been found under the bark of waterlogged branches of various deciduous trees and in sap runs on *Quercus* (*S. clunipes* by Rotheray, 1990) and *Ulmus* (*S. elegans* and *S. verecunda*, by Hartley, 1961), see Speight (2003).

Recognition

Sphegina are slender, elongate flies. Their abdomen is claviform with segment 2 elongate and segments 3 and 4 broadened, a feature especially prominent in the female. Their colour is usually dark, even blackish. Some species have large reddish markings on tergites 3 and 4. *S. sibirica* has a yellow colour form. The key is based on Thompson and Torp (1986), Doczkal (1995), Vujic (1990) and Kassebeer (1991).

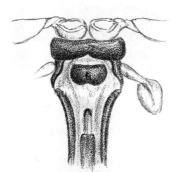

figure 732. *Sphegina clunipes*, thorax from below (Verlinden).

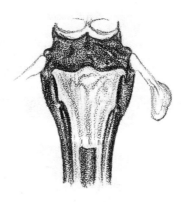

figure 733. *Sphegina sibirica*, thorax from below (Verlinden).

Key

1.a. First sternum not reduced, wider than long, oval (figure 732 **❯** 2

1.b. First sternum greatly reduced (and longer than wide) or absent (figure 733); face and humerus black; katepisternum shiny. 7-8 mm. Northern, Western and Central Europe, Siberia, extending its range **❯ *Sphegina sibirica*** Stackelberg
Jizz: specimens black to yellow (var. *flavescens*) with all intermediates.

2.a. Katepisternum dull, grey pollinose, at most upper margin shiny **❯** 3

2.b. Katepisternum shiny, without pollinosity; hind coxa black; face usually entirely black. 5-6 mm. Northern and Central Europe **❯ *Sphegina montana*** Becker

3.a. Tibia 2: not expanded (figure 749), male: tibia 2 not expanded, if some expansion appears then legs 1 and 2 extensively dark; female: tibia 2 of equal width over its length, at apex at most with some brownish bristles **❯** 4

3.b. Tibia 2 expanded on distal 1/2, male: greatly expanded on apical 1/4 (figure 734), female: expanded on apical 1/2 and with dense black bristles at the apex. 7-9 mm. Elevated parts of Central Europe **❯ *Sphegina platychira*** Szilady

4.a. Males (eyes separated, check for genitalia) **❯** 5

4.b. Females (females of *S. claviventris* not known) **❯** 13

5.a. Face entirely dark brownish to black (figure 735) **❯** 6

5.b. Face pale yellow to orange on ventral 1/4 or more (figure 736, figure 737) **❯** 8

6.a. Hind coxa partially pale, yellow to orange **❯** 7

6.b. Hind coxa dark, brownish to black; frons with long black hairs, with some hairs longer than 3rd antennal segment width. 6-7 mm. Northern Europe and mountains of Central Europe, Siberia **❯ *Sphegina spheginea*** (Zetterstedt)

7.a. Antenna pale orange; subcosta joining costa before r–rm crossvein (figure 738); surstyle narrowed and pointed apically (figure 740). 5-7 mm. Central Europe **❯ *Sphegina nigra*** Meigen (= *Sphegina clavata Scopoli*)

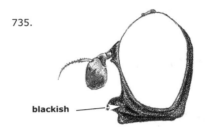

735.

blackish

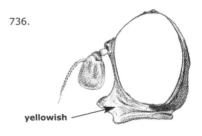

736.

yellowish

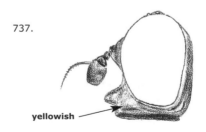

737.

yellowish

figure 734. *Sphegina platychira*, leg 2 (after Van der Goot, 1981).

figure 735. *Sphegina verecunda*, head of male.
figure 736. *Sphegina elegans*, head of male.
figure 737. *Sphegina clunipes*, head of male (Verlinden).

7.b. Antenna darker, brownish-black to black on basal segments, 3rd segment may be paler brown; subcosta joining costa at or beyond r–m crossvein (figure 739); surstyle broad, blunt apically (figure 741). 5-6 mm. Central Europe ❯ *Sphegina verecunda* Collin

8.a. Hind coxa and trochanter brownish-black to black; genitalia enlarged; with surstyle elongate and always visible externally; 4th sternum expanded with apicolateral tufts of long bristles (figure 742) ❯ 9

8.b. Hind coxa and trochanter partially yellow; genitalia not enlarged; with surstyle shorter and concealed under 4th sternum; 4th sternum flat and without bristles ❯ 11

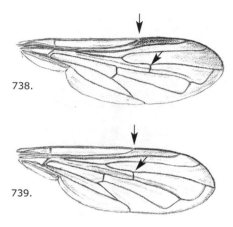

738.

739.

figure 738. *Sphegina clavata*, wing.
figure 739. *Sphegina verecunda*, wing (Verlinden).

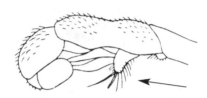

figure 742. *Sphegina latifrons*, tip of male abdomen (after Van der Goot, 1981).

740.

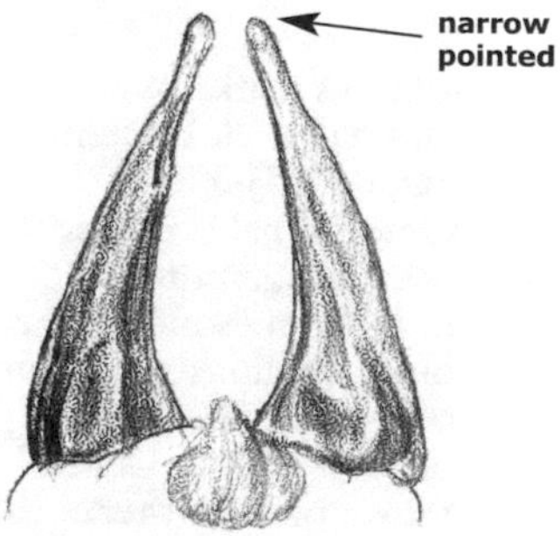

narrow pointed

741.

broad blunt

743.

744.

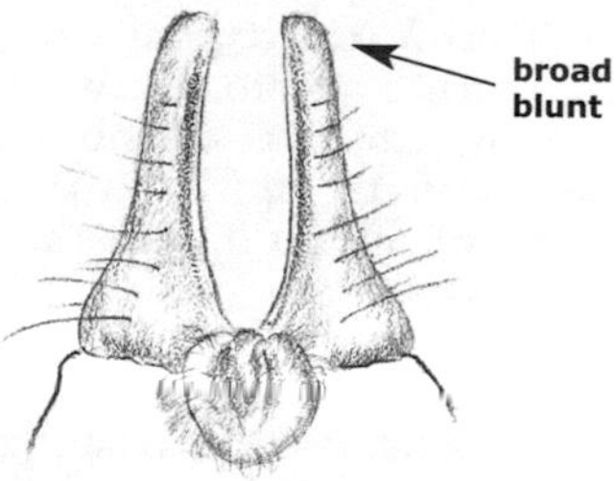

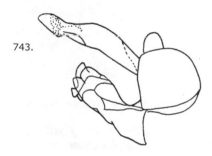

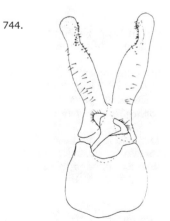

figure 740. *Sphegina clavata*, genitalia of male.
figure 741. *Sphegina verecunda*, genitalia of male (Verlinden).

figure 743. *Sphegina latifrons*, genitalia of male. (Vujic, 1990)
figure 744. *Sphegina latifrons*, genitalia of male (Thompson and Thorp, 1986).

745.

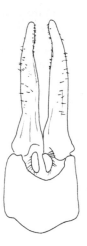

747.

748.

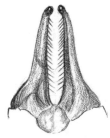

746.

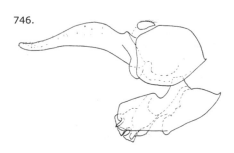

figure 747. *Sphegina claviventris*, genitalia of male. (Thompson and Thorp. 1986)
figure 748. *Sphegina clunipes*, genitalia of male (Verlinden).

figure 745. *Sphegina cornifera*, genitalia of male.
figure 746. *Sphegina cornifera*, genitalia of male (Thompson and Thorp, 1986).

9.a. Surstylus broadly rounded apically (figure 743, figure 744); sternite 3 wider than long **>** 10
9.b. Surstylus acute apically (figure 745, figure 746); sternite 3 longer than wide. 7-8 mm. Elevated parts of Central Europe **>** *Sphegina cornifera* Becker

10.a. Bare longitudinal stripe on frons as wide as ocellular triangle; tergite 1 medially dusted for 1/3; face yellow from mouth edge up to the lower border of eye. 6-7 mm. Central Europe **>** *Sphegina latifrons* Egger
10.b. Bare longitudinal stripe on frons much smaller than ocellular triangle; tergite 1 medially dusted for 3/4; only the front part of mouth edge yellow. 6-7 mm. Central France to Pyrenees **>** *Sphegina varifacies* Kassebeer

11.a. Humerus dark, brownish-black to black; antennae: 3rd antennal segment small (figure 737) **>** 12
11.b. Humerus pale, yellow to orange; antennae: 3rd antennal segment large (figure 736); mesonotum extensively shiny black, pollinose only on margins; front and middle tarsi entirely pale; wing hyaline. 6-7 mm. Central Europe, mountains of Southern Europe **>** *Sphegina elegans* Schummel

12.a. Tibia 3 transverse apically; 4th tergite short, much broader than long; genitalia asymmetrical (figure 747). Large species: 10-11 mm. Northern Europe, Northern Siberia **>** *Sphegina claviventris* Stackelberg
12.b. Tibia 3 with a small apicoventral spur; 4th tergite long, about as broad as long; genitalia symmetrical (figure 748); front and middle tarsae usually entirely pale yellow, rarely with apical segments darker brownish-orange. 6-7 mm. Northern and Central Europe, mountains of Southern Europe, in Asia to Pacific coast **>** *Sphegina clunipes* Fallén

13.a. Tergite 2: longer than wide at hind margin; frons: hairs short, as long as the distance between the hind ocelli; thoracic dorsum with pale hairs only; legs 1 and 2: pale, at most with darkened tarsi and a brown ring on tibia; femur 3: black over more than 1/2 its length; alula narrow **›** 14

13.b. Tergite 2: shorter than wide at hind margin; frons: hairs long, almost as long as 1/2 the frons width, partly black; thoracic dorsum partly black-haired; legs 1 and 2 brown; femur 3: black in apical 2/5; alula wide **›** *Sphegina spheginea* Zetterstedt

14.a. Tergite 5: distal 1/2 with 2 parallel membraneous strips, due to incisions of the hind margin; mouth edge strongly protruding; genae wide **›** 15

14.b. Tergite 5 without membraneous strips; mouth edge moderately protruding; genae small **›** 17

15.a. Frons with bare longitudinal stripe; cerci long (length:width = 2:5 or more) **›** 16

15.b. Frons without bare longitudinal stripe, completely haired; cerci short (length:width = 2:2) **›** *Sphegina cornifera* Becker

16.a. Face yellow in lower parts **›** *Sphegina latifrons* Egger

16.b. Face black **›** *Sphegina varifacies* Kassebeer

17.a. Postpronotum and humeri brown to black **›** 18

17.b. Postpronotum and humeri yellow **›** *Sphegina elegans* Schummel

18.a. Face: black, only in rare occasions with little yellow in lower 1/2; tergite 2: at most 2.5 times as long as wide at front margin, front margin with dusted stripe; frons: with a clear bare stripe, sometimes narrowed in the middle **›** 19

18.b. Face: extensively yellow in lower parts; tergite 2: more than 2.5 times as long as wide at front margin, front margin mostly undusted; frons: without or only with a smallish bare stripe **›** *Sphegina clunipes* Fallén

19.a. Subcosta joining costa before r-rm crossvein (figure 738); sternite 1 longer, with posteromedian excision; frons: bare stripe small: at the front ocellus as wide as the diameter of an ocellus **›** *Sphegina nigra* Meigen (= *Sphegina clavata* Scopoli)

19.b. Subcosta joining costa at r-rm crossvein (figure 739); sternite 1 shorter, without posteromedian excision; frons: bare stripe broad: at the front ocellus as wide as the ocellular triangle **›** *Sphegina verecunda* Collin

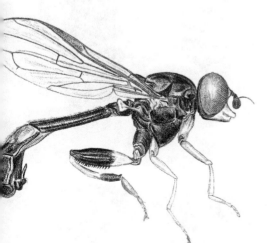

figure 749. *Sphegina clunipes*, habitus of male (Verlinden).

SPILOMYIA

Introduction

Spilomyia hoverflies are convincing social wasp mimics in their colour, form and behaviour. They are reported to wave their front legs to imitate antennae, just as *Temnostoma* species do. They are large, but they tend to 'disappear into the crowd' when many social wasps are present because of their excellent mimicry. The flies are encountered in various situations: flying near vegetation, visiting flowers or basking in the sun on stones. The flies also inspect trees and rot holes, their larval habitats.

Recognition

The typical characteristics of this genus are the short antennae (differentiating it from *Chrysotoxum*), the striped eyes and the conical tooth on femur 3. The thoracic dorsum has a characteristic V-shaped marking in front of the scutellum. The key is based on Barendregt et. al., Van Steenis and Van Steenis (2000) and Van Steenis (2000).

Key

1.a. Pleura black with more than 3 yellow spots. Posterior margin of the scutellum yellow. Hairs on thoracic dorsum, scutellum and pleura ranging from medium length to short and adpressed. Legs yellow-brown and black ❭ 2

1.b. Pleura black with 3 yellow spots (figure 750). Posterior margin of the scutellum orange-red. Hairs on thoracic dorsum, scutellum and pleura long and erect. Legs yellow-brown to red. Abdomen predominantly black, with narrow yellow bands. 12-16 mm. Central Europe, Siberia ❭ ***Spilomyia diophthalma*** Linneaus

2.a. Thoracic dorsum with short, often adpressed hairs. Pleura with 4, 5 or 6 yellow spots, katatergite always yellow. Front tars black, but often with apical segments yellow; front tibia black over less than their apical 1/2, occasionally entirely yellow ❭ 3

2.b. Thoracic dorsum with fairly long, erect hairs. Pleura normally with 5 yellow spots (figure 751), if only 4 spots present, then the katatergite (location of hind spot) black. Front tars black, front tibia black over more than its apical 1/2. Abdomen with the anterior yellow band on tergites 2 and 3 uninterrupted or slightly separated in the middle. 11-16 mm. Central and Southern Europe ❭ ***Spilomyia manicata*** Rondani (= *Spilomyia boschmai* Lucas)

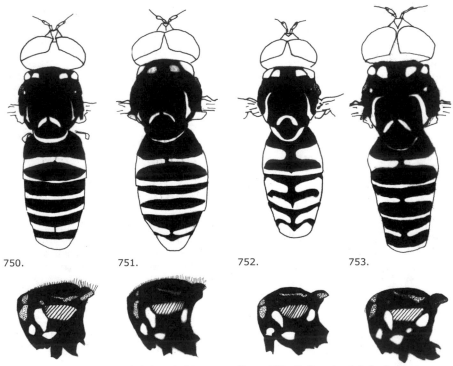

750.   751.   752.   753.

figure 750. *Spilomyia diophthalma*, habitus.
figure 751. *Spilomyia manicata*, habitus (Wakkie).

figure 752. *Spilomyia digitata*, habitus.
figure 753. *Spilomyia saltuum*, habitus (Wakkie).

3.a. At least apical 1/4 of front tibia black, at most 5th segment of front tarsus yellow; bristles on front tarsus all black; pleura with 4 yellow spots; abdominal bands orange-yellow, with the posterior bands on tergites 3 and 4 strongly curved (figure 752). 10-16 mm. Southern Europe, extinct in Central Europe ❯ *Spilomyia digitata* Rondani

3.b. Front tibiae yellow, at most with a black spot on apical 1/8. At least 4th and 5th segments of front tarsus yellow; bristles on at least ventral side of 4th and 5th segments of front tarsus yellow. Pleura with 5 yellow spots. Abdomen with narrow yellow bands, the anteromedial yellow band on tergites 2-4 slightly separated in the middle (figure 753). 10-15 mm. Central and Southern Europe, Turkey ❯ *Spilomyia saltuum* Fabricius

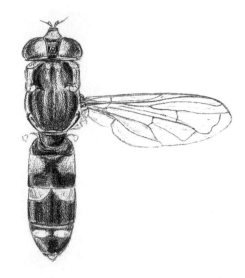

figure 755. *Syritta pipiens*, habitus of female (Verlinden).

SYRITTA

Introduction

Syritta pipiens is a very common hoverfly found everywhere that plants are flowering. They are small, linear hoverflies with a characteristic flight pattern; they hover, make a small turn and a quick flight, hover again, etc. Their larvae live in decaying plant material.

Key

1. Male: femur 3 strongly thickened, but hardly bent, basally without protuberance; tergites 2 and 3 with small, pale spots (figure 754). Female: ocellar triangle black or bluish, metallic sheen; thoracic dorsum dusted along side margin; tergite 4: side and hind margins not dusted (figure 755). 7-9 mm. Cosmopolitan ❯ *Syritta pipiens* Linnaeus
Jizz: small 'stick' hovering in front of flowers or in vegetation.

SYRPHUS

Introduction

Syrphus are common black and yellow hoverflies in the garden and along forest edges. They inhabit a variety of habitats, but prefer woodlands and shrubs. Males can be found hovering 1-5 metres above the ground, apparently guarding an air space. The larvae of *Syrphus* are aphid predators (figure 756).

figure 754. *Syritta pipiens*, habitus of male (Verlinden).

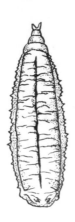

figure 756. Larva of Syrphus torvus

figure 757. *Syrphus vitripennis*, habitus of male (Verlinden).

Recognition

Syrphus contain broad-bodied, black and yellow flies with a pollinose, greenish-black thorax. The face never has a black facial stripe, but the mouth edge may be black. They are most easily confused with *Parasyrphus* species. Most *Parasyrphus* are smaller and have a black facial stripe (except for two species). In doubtful cases, check the anterior anepisternum for hairs (*Parasyrphus* has hairs there, *Syrphus* does not), and check the lower lobe of the calypter for hairs (only *Syrphus* has hairs on the upper side). The key is based on Barendregt (1983), Vockeroth (1992) and Goeldlin de Tiefenau (1996).

Key

1.a. Wing with 2nd basal cell (cell bm) bare on anterior 1/4 or more; upper surface of the tip of femur 3 with short adpressed yellow hairs, sometimes mixed with some black hairs ❭ 2
1.b. Wing entirely trichose; upper surface of the tip of femur 3 with short adpressed black hairs ❭ 4

2.a. Frons immediately above lunulae yellow; tergites with paired yellow spots; lateral margin of tergites yellow; face yellow. 8–10 mm. Northern Europe, Northern Siberia, Nearctic ❭ *Syrphus sexmaculatus* Zetterstedt
2.b. Frons immediately above lunulae black; tergites mostly with entire yellow bands; lateral margin of tergites black ❭ 3

3.a. Female: femur 3 black on about basal 2/3. Males indistinguishable from next (figure 757). 8–11 mm. ❭ *Syrphus vitripennis* Meigen
3.b. Female: femur 3 yellow on basal 1/2, usually partly brownish on apical 1/2. Males indistinguishable from previous. 8–11 mm. North American species, found at several sites in Central Europe, but status doubtful (Speight, 2003) ❭ *Syrphus rectus* Osten Sacken
Note: Bartsch et al. (2009a) doubs the presence of the North American *S. rectus* in Northern Europe because they argue it falls within the range of variation of *S. vitripennis*.

4.a. Eyes with numerous hairs, in male long and dense, in female shorter; femur 3 black on basal 3/4. 10-13 mm. Europe, in Asia to Japan, Nearctic ❭ *Syrphus torvus* Osten Sacken
4.b. Eye with at most some very short scattered hairs. Female: femora yellow, except in *S. nitidifrons* with its black femur with yellow tip ❭ 5

5.a. Mouth edge yellow, at most genae and posterior mouth edge narrowly black; frons yellow or yellow and black, dusted; tibia 3 yellow; femur 1 with long white to yellow hairs. Female: femur 3 yellow ❭ 6
5.b. Mouth edge black (figure 758); frons shiny black; tibia 3 mainly black; femur 1 with long black hairs; tergites with spots. Female: femur 3 black with yellow apex; habitus (figure 759). 7-10 mm. Central Europe ❭ *Syrphus nitidifrons* Becker

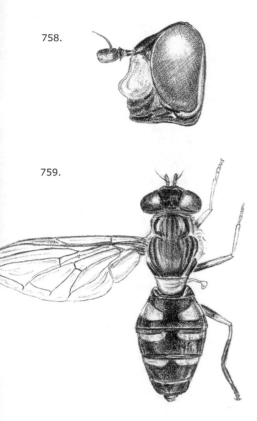

758.

759.

figure 758. *Syrphus nitidifrons*, head of male.
figure 759. *Syrphus nitidifrons*, habitus of
female (Verlinden).

6.a. Metatars 2: a portion of the bristles
on the ventral surface black. Male: eyes
meeting above the antennae for a distance longer than that between the
anterior and posterior ocelli **〉** 7

6.b. Metatars 2: all bristles and hairs on
the ventral surface orange-yellow. Male:
eyes meeting on frons for a distance
equal to or longer than the distance
between the anterior and posterior
ocelli. 8-11 mm. Alps, may also occur at
lower altitudes. **〉 *Syrphus auberti***
Goeldlin

7.a. Frons, immediately posterior to the
lunulae, yellow; sternites yellow. Male:
femur 3 yellow. Female: frons shiny yellow and entirely covered with silver-
grey dusting for all but the posterior 1/4
of its length; posterior 1/4 of the frons,
including the ocellar triangle, shiny
black **〉** 8

7.b. Frons, immediately posterior to the
lunulae, shiny black; sternites frequently
black marked laterally and medially.
Male: femur 3 black for 2/3 of its length;
female: black area on frons with a pos-
tero-median triangular extension pointing towards the occiput, then entirely
covered with silver-grey dusting back
almost as far as the ocellar triangle, ocellar triangle and posterior 1/4 of the frons
shiny black; femur 3 yellow; lateral margins of tergites yellow only at the ends of
the bands (rarely paired spots); antennae
black above, orange below; front tarsus
darkened. 10-12 mm. Europe, in Asia
into Japan, Nearctic highly migratory **〉**
Syrphus ribesii (Linnaeus)

8.a. Lateral margins of tergites continuously yellow. Male: legs yellow, at most the
very base of femora 1 and 2 black and
femur 3 black on its basal 1/4. Female:
legs entirely yellow; antennae usually distinctly orange with 3rd segment somewhat darkened above. 8-11 mm.
Northern Europe, Northern Siberia,
Nearctic **〉 *Syrphus attenuatus*** Hine

8.b. Lateral margins of tergites partly
black and partly yellow. Male: femora 1
and 2 black on basal 1/3 of their length,
femur 3 black for 2/3 of its length.
Female: legs yellow with all tarsi darkened, tarsus 3 almost black. 8-11 mm.
Northern Europe **〉 *Syrphus admiran-
dus*** Goeldlin

TEMNOSTOMA

Introduction

Temnostoma dwell in damp deciduous
forests, where they bask on leaves and
twigs and visit the flowers of understorey
herbs. They exhibit territorial behaviour,
chasing any larger insects in the immediate vicinity. The larvae of *Temnostoma* live
in old dead wood of deciduous trees.
Derksen (1941) indicates that metamor-
phosis of *T. bombylans* takes two years and
the larvae inhabit stumps of trees felled
seven to eight years previously.

Recognition

Temnostoma are convincing wasp mimics. Flying *T. vespiforme* and *T. apiforme* even mimic the erratic flight patterns of wasps. However, once sitting they are quickly recognisable as hoverflies by their short antennae and unfolded wings. The hoverfly genus *Sphecomyia* is similar to *Temnostoma* but has long antennae. *Spilomyia* is superficially like *Temnostoma* but has striped eyes and two yellow stripes on the thoracic dorsum before the scutellum. Key based on Krivosheina (2004) and Bartsch et al. (2009b).

Key

1.a. Thoracical dorsum with triangular yellow spots just before the postalar lobes (figure 760) **>** 2
1.b. Thoracical dorsum black just before postalar lobes (figure 761) **>** 6

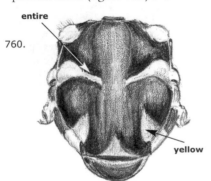

760.

entire

yellow

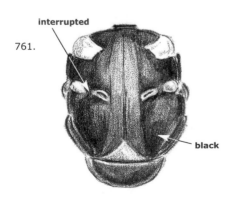

interrupted

761.

black

figure 760. *Temnostoma vespiforme*, thoracic dorsum.
figure 761. *Temnostoma apiforme*, thoracic dorsum (Verlinden).

2.a. Tergite 3-4 and often tergite 2 with a pair of yellow stripes **>** 3
2.b. Tergite 2-4 with a single yellow stripe, at the hind margin at most a small stripe of grey dust without clear margin. Northern Europe, Siberia. **> *Temnostoma sericomyiaeforme*** Portschinsky

3.a. Tergite 2-4: front and hind yellow band broadly connected at the side margin, the width of the black band in between 2-2.5 times smaller than the width of the anterior yellow band. **>** 4
3.b. Tergite 2-4: front and hind yellow band separated at the side margin, at most joined by a small yellow strip. **>** 5

4.a. Thoracic dorsum: the yellow spot before the postalar lobes reaches and covers the postalar lobes, which are partly yellow; the yellow streak at suture of the thoracic dorsum uninterrupted. 14-17 mm. Northern and central Europe, in Asia into Japan, Nearctic region **> *Temnostoma vespiforme*** Linnaeus
4.b. Thoracic dorsum: the yellow spot before the postalar lobes does not reach the postalar lobes, which are black; the yellow streak at the suture interrupted in the middle. 14-17 mm. Northern and Central Europe **> *Temnostoma meridionale*** Krivosheina and Mamaev

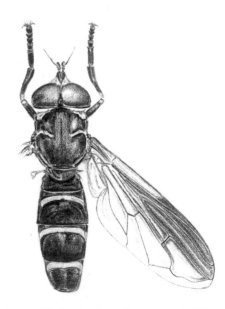

figure 762. *Temnostoma bombylans*, habitus of male (Verlinden).

5.a. Tergite 3-4: central, dark band more than twice as wide as the anterior yellow band; tergite 3: yellow band at hind margin narrow: its width 1/3-1/2 times the width of the front yellow stripe. Siberia. ❭ **Temnostoma sibiricum** Portschinsky

5.b. Tergite 3-4: central, dark band as wide as the anterior yellow band; tergite 3: yellow band at hind margin wider, at least 2/3 of the width of the front yellow band, often wider. ❭ see 4.a. *Temnostoma vespiforme* Linnaeus

Note: two dark subspecies exist in Siberia, ssp. *tuwensis* and *altaicum* Krivosheina, 2004

6.a. Postalar knobs with short, adpressed to semi-erect hairs; abdomen: tergite 3-4 with 1 yellow band, always black at hind margin. Male: thoracic dorsum longer than wide, rectangular ❭ 7

Jizz: rather elongate species, abdomen almost linear.. Wings heavily infuscate, the dark spot reaches the front in the top half.

6.b. Postalar knobs with erect, long hairs; abdomen: tergite 3-4 either with 1 band and hind margin of tergite 4 with narrow grey dust or with 2 bands, one on front half and one along hind margin. Male: thoracic dorsum approximately square ❭ 8

Jizz: rather broad species, abdomen more ovoid. Wings less infuscate, often the front of the wing more or less hyaline.

7.a. Tibia 3: either entirely yellow or with a vague black streak in the apical half. Female: frons with yellow dust band along eye margin. 12-14 mm. Europe, North Africa, in Asia into Japan ❭ *Temnostoma bombylans* Fabricius

7.b. Tibia 3: black for one third or more of its length, black often fading on inner side. Female: frons with narrow silvery dust band. Northern Europe, Siberia ❭ *Temnostoma angustistriatum* Krivosheina

8.a. Abdomen: Tergite 2-4(-5 for females) with 2 distinct yellow bands. Females: width of dust spots on frons 1/6-1/2 of the undusted stripe in between. 13-15 mm. Northern Europe and northern part and mountains of middle Europe, in Asia into Japan ❭*Temnostoma apiforme* Fabricius ❭ 9

8.b. Tergite 2-4(-5 for females) with 1 distinct yellow band and very narrow grey dust at the hind margin of tergite 4 and 5. Female: frons with dust spots narrow, their width 1/7 to 1/8 of the width of the undusted stripe in between. ❭ *Temnostoma carens* Gaunitz

TRICHOPSOMYIA

Introduction

Trichopsomyia (= *Parapenium*) are small black hoverflies that occur in damp forests. They resemble a small *Pipiza*. Females of *T. lucida* and *T. flavitarse* have yellow spots on tergite 2, those of *T. joratensis* are black. The key is based on Goeldlin (1997) and Barendregt ('determinatiemap').

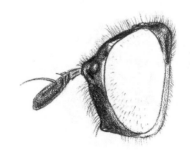

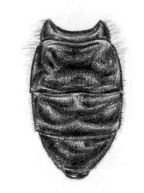

figure 764. *Trichopsomyia joratensis*, abdomen of female (Verlinden).

765.

766.

figure 765. *Trichopsomyia flavitarse*, wing tip.
figure 766. *Trichopsomyia lucida*, wing tip
(Verlinden).

767.

768.

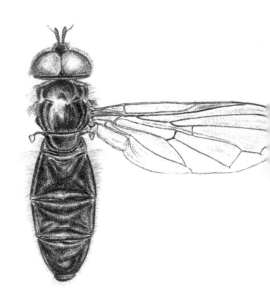

figure 767. *Trichopsomyia flavitarse*, genitalia of male.
figure 768. *Trichopsomyia lucida*, genitalia of
male (Verlinden).

769.

770.

figure 769. *Trichopsomyia flavitarse*,
abdomen of female.
figure 770. *Trichopsomyia lucida*, abdomen
of female (Verlinden).

figure 771. *Trichopsomyia flavitarsis*, habitus
of male (Verlinden).

Key

1.a. Antennae: 3rd segment elongated, at least 2-3 times as long as wide (figure 763); face, in front view, smaller than an eye. Male: genitalia: surstylus narrowed in top 1/2, pointed at tip. Female: tergite 2 with 2 yellow spots which may be reduced; occiput: lower part greyish dusted; tars 2: metatars pale, other segments dark ❭ 2

Note: males of *lucida* can be quite large (pers. comm. Barendregt).

1.b. Antennae: 3rd segment shorter, at most 1.5 times as long as wide; face, in front view, 1.5 times as wide as an eye. Male: genitalia: surstylus very broad, not narrowed in top 1/2, thick and broadly rounded at tip. Female: tergite 2 black (figure 764); frons without dust spots; tars 2: first 2 segments pale, others dark. Larger species: 7-8 mm. Northern and Central Europe ❭ **Trichopsomyia joratensis** Goeldlin de Tiefenau (= *Trichopsomyia carbonaria* auctorum)

2.a. Wing: vein tm ends (almost) perpendicular to R4+5 (figure 765), in males with a slightly sharp angle (figure 771); abdomen shiny. Male: 3rd antennal segment 3 times as long as broad. Female: tergite 2 with 2 relatively small, round spots (figure 769); occiput with dense silverish hairs; frons: without dust spots. 5-6 mm. Northern and Central Europe, in Asia to Pacific coast ❭ **Trichopsomyia flavitarsis** Meigen

Jizz: dull species.

2.b. Wing: vein tm ends in a sharp angle on R4+5 in the wing, in males with a very sharp angle (figure 766); abdomen dull by chagrination. Male: 3rd antennal segment twice as long as broad. Female: tergite 2 with 2 relatively large, squarish to triangular spots (figure 770); occiput with sparse hairs; frons: with a pair of triangular silver-grey dust spots. 5-6 (8) mm. Central Europe ❭ **Trichopsomyia lucida** Meigen

Jizz: shiny species.

TRIGLYPHUS

Introduction

Triglyphus are very small hoverflies with long wings, which live in herbaceous plants.

Key

1. Small, black species (figure 772, figure 773), only knees and metatars 2 pale; abdomen: tergites 2 and 3 large, tergite 4 very small and inconspicuous. 5-6 mm. Central and Southern Europe, in Asia into Korea ❭ **Triglyphus primus** Loew

Jizz: a cross between a *Pipiza* and a *Paragus* in the field, also reminiscent of a Pipunculidae because of its long wings.

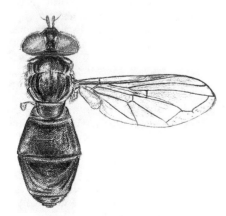

figure 772. *Triglyphus primus*, habitus of female (Verlinden).

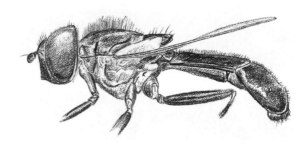

figure 773. *Triglyphus primus*, habitus of male (Verlinden).

774.

deeply curved

775.

gently curved

figure 774. *Tropidia fasciata*, wing.
figure 775. *Tropidia scita*, wing (Verlinden).

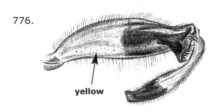

776.

yellow

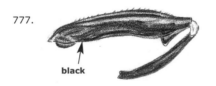

777.

black

figure 776. *Tropidia fasciata*, femur 3 of female.
figure 777. *Tropidia scita*, femur 3 of female (Verlinden).

figure 778. *Tropidia scita*, habitus of male (Verlinden).

TROPIDIA

Introduction

Tropidia inhabit wet habitats near forests with running or standing water, such as wetlands, streams and ditches, without tree cover in the immediate vicinity. Adults are found in herbaceous vegetation, where they dart erratically between stems or quickly fly over the vegetation. Males of *T. scita* repeatedly hover over the vegetation, exposing the red pattern on their abdomens, often in front of other males. The larvae are presumed to live in the debris layer of waterways. *T. fasciata* has been found ovipositing on a stream edge, larvae of *T. scita* have been found associated with cattails (*Typha*) (Speight, 2003).

Recognition

Tropidia are elongate hoverflies with a rather large thorax, reminiscent of *Xylota*. Unlike *Xylota*, *Tropidia* have femur 3 very swollen with a protuberance at the tip and they fly more than they walk. The red pattern on the abdomen is interrupted with a median black part, unlike the *Xylota* species with a red pattern on the abdomen.

Key

1.a. Wing: R4+5 bent in underlying cell (figure 774); thoracic dorsum bronze, weakly shiny; antennae: reddish yellow, 3rd segment may be darkened on upper margin. Female: femur 3 pale on basal 1/2 (figure 776). 10-11 mm. Northern and Central Europe, Eastern Siberia **>** ***Tropidia fasciata*** Meiger
Jizz: behaves just like a *Merodon*, often near streams.

1.b. Wing: R4+5 weakly bent into the underlying cell (figure 775); thoracic dorsum shiny black beyond dusted parts; antennae: brown to black, seldom turning yellow; habitus figure 778. Female: femur 3 black, pale on tip (figure 777). 9-10 mm. Northern and Central Europe, in Asia into Japan **>** ***Tropidia scita*** Harris
Jizz: a wetland '*Xylota*' with orange spots and swollen femur 3, may be numerous in *Iris pseudacoris* flowers.

VOLUCELLA

Introduction

Volucella are encountered in and along the edges of deciduous forests, where the males hover at several metres above the ground. *V. zonaria* is associated with urban areas in The Netherlands, with many sightings on *Buddleia* in gardens. It is a strongly migratory species, and probably tries to establish populations north of the range where it is able to hibernate each year. Adult *Volucella* forage on umbelliferae and several flowering shrubs (e.g. *Crataegus*, *Ligustrum*). They are also found sunning themselves on leaves.

The larvae live in the nests of bumblebees and social wasps, where they are detritivores and larval predators. The exception is probably *V. inflata*, whose larvae appear to be inhabitants of tunnels made by other insects in which sap and insect faeces/tree humus provide a sub-aqueous mix (Speight, 2003).

Recognition

Volucella are large, broad-bodied hoverflies. The face is extended downwards and the arista is densely feathered. Cell R1 in the wing is closed before the wing border.

Key

1.a. Thorax and scutellum with short, adpressed hairs; abdomen black and yellow or black and white; scutellum with bristles at hind margin **>** 2

1.b. Thorax and abdomen with long dense hairs, bumblebee mimic; abdomen black, seldom with orange spots on tergite 2; scutellum without bristles at hind margin. 11-15 mm. Europe, in Asia to Japan, Nearctic **>** ***Volucella bombylans*** Linnaeus

Jizz: broad and hairy, scutellum black; a number of colour forms are recognized;

a. typical. Hairs on thorax and first part of abdomen black, tip of abdomen with reddish hairs;

b. *plumata*. Hairs on thorax yellowish, mostly with black hairs on the part behind the head, hairs on base of abdomen yellow, on its tip white, with black hairs in between;

c. *haemorrhoidalis*. As *plumata*, but tip of abdomen with reddish hairs.

2.a. Tergites: yellow with black bands (figure 779) **>** 3

2.b. Tergites: tergite 2 with yellow or white markings, tergites 3 and 4 black (figure 780) **>** 4

figure 779. *Volucella inanis*, habitus of male (Verlinden).

figure 780. *Volucella pellucens*, habitus of male (Verlinden).

3.a. Thoracic dorsum reddish-yellow, shiny; tergites shiny, yellow, with dark bands at the hind margin of tergites 2 and 3; sternite 2 black with yellow hind margin (figure 781). 18-20 mm. Central and Southern Europe, North Africa, in Asia into Japan, highly migratory **>** *Volucella zonaria* Poda

Jizz: large, broad and yellow and red, with striking yellow face.

3.b. Thoracic dorsum blackish, dull; tergites dull, yellow, with dark bands at the hind margin of tergites 2, 3 and (often) 4; sternite 2 yellow (figure 782). 15-16 mm. Europe, in Asia to Pacific coast **>** *Volucella inanis* Linnaeus

Jizz: smaller, dull *zonaria*, thoracic dorsum blackish.

4.a. Thoracic dorsum black, at the side margin turning brownish; thoracic hairs mostly black; tergite 2 with white spots, which may turn yellowish. Male: mouth edge not elongated downwards (figure 783). Female: eyes narrowly separated (figure 785). 13-18 mm. Europe, in Asia into Japan, Oriental region **>** *Volucella pellucens* Linnaeus

783.

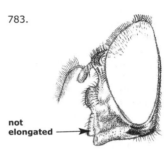

not
elongated

781.

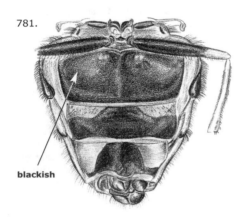

blackish

784.

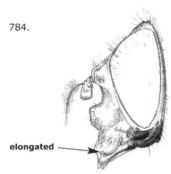

elongated

figure 783. *Volucella pellucens*, head of male.
figure 784. *Volucella inflata*, head of male (Verlinden).

782.

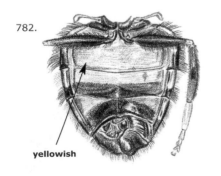

yellowish

785.

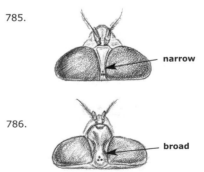

narrow

786.

broad

figure 781. *Volucella zonaria*, abdomen from below.
figure 782. *Volucella inanis*, abdomen from below (Verlinden).

figure 785. *Volucella pellucens*, head of female, top view.
figure 786. *Volucella inflata*, head of female, top view (Verlinden).

4.b. Thoracic dorsum black, at the side margin yellowish; thoracic hairs brown to yellow; tergite 2 with yellow spots. Male: mouth edge strongly elongated downwards (figure 784). Female: eyes broadly separated (figure 786). 13-16 mm. Central Europe **>** *Volucella inflata* Fabricius

XANTHANDRUS

Introduction

Xanthandrus dwell in deciduous and coniferous forests, especially those with a thick understorey of shrubs. Males hover at 3-5 metres. The larvae are known to prey on aphids and the caterpillars of various small moths (e.g. Tortricidae), both on trees and low-growing plants. They are also known predators of the caterpillars of the pine procession moths (*Thaumetopoea pinivora* and *T. pityocampa*) (see Rotheray, 1993; Speight, 2003).

figure 787. *Xanthandrus comtus*, habitus of male (Verlinden).

Recognition

Broad-bodied hoverflies, with a black scutellum and face and reddish-orange markings on the abdomen.

Key

1. Only one European species. Abdomen broad, with yellow spots: male with round spots on tergite 2 and confluent spots on tergites 3 and 4 (figure 787), female with oval spots on tergite 2 and square spots on tergites 3 and 4; wing with long, black pterostigma; 10-12 mm. Europe, in Asia into Japan **>** *Xanthandrus comtus* Harris
Jizz: large '*Melanostoma*' with broad abdomen.

XANTHOGRAMMA

Introduction

Xanthogramma live along forest edges and shrubby vegetation. They regularly occur in urban areas, especially those with well-established gardens. Their larvae are aphid predators.

Recognition

Large, yellow and black hoverflies, with a strong contrast between the yellow and the black. Here, the former genus *Olbiosyrphus* is included in *Xanthogramma*. Additional species are under discussion. Large, yellow and black hoverflies, with a strong contrast between the yellow and the black. Here, the former genus Olbiosyrphus is included in Xanthogramma. Key based on Van der Goot (1981) and Bartsch et al. (2009a). *Xanthogramma dives* (Rondani) is also mentioned from Northwest Europe. Because it is inadequately separated from X. *pedissequum* and X. *stackelbergi*, it is not included in the key.

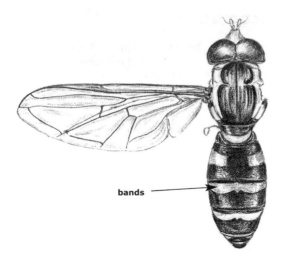

figure 788. *Xanthogramma laetum*, habitus of male (Verlinden).

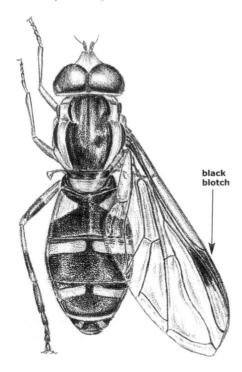

figure 789. *Xanthogramma pedissiquum*, habitus of male (Verlinden).

Key

1.a. Eyes bare; wing with dark blotch at tip; tergites 3 and 4 with elongate spots. 10-12 mm ❯ 2

1.b. Eyes haired; wing hyaline; tergites 3 and 4 with bands, which may be just interrupted in the male. 8-10 mm. Central Europe ❯ *Xanthogramma (Olbiosyrphus) laetum* Fabricius

2.a. Femur and tibia 3 pale with dark ring; tergite 2 with broader spots than those on tergites 3 and 4, spots on tergite 2 triangular (figure 789) ❯ 3

2.b. Femur and tibia 3 completely pale; tergites 2-4 with subequal spots, spots on tergite 2 linear (figure 790). 10-12 mm. Central and Southern Europe, Western Siberia ❯ *Xantogramma citrofasciatum* DeGeer (= *X. festivum* Linnaeus)

3.a. Wing: the dark patch around the stigma extends to two cells below the stigma; abdomen viewed from below: the membrane between the sternites and tergites at each segment yellow in front part and black in hind part. 10-13mm. Europe, Western Siberia ❯ *Xanthogramma pedissequum* Harris

3.b. Wing: the dark patch small, extends only to the cell just below the stigma; abdomen viewed from below: the membrane between the sternites and tergites entirely yellow on segments 3-5. 10-12mm. Northern Europe, Siberia. ❯ *Xanthogramma stackelbergi* Violovitsh

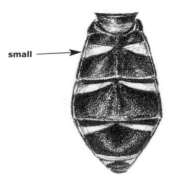

figure 790. *Xanthogramma citrofasciatum*, abdomen of male (Verlinden).

XYLOTA

Introduction

Xylota wander around on leaves on the forest edge. Species like *X. segnis* feed on the pollen deposited on leaf surfaces. Other species, like *X. caeruleiventris*, regularly visit flowers.

The larvae of *Xylota* are saprophagous and live in sap runs of trees, under bark and in decaying heartwood. *X. segnis* larvae are also found in decaying vegetation (Rotheray, 1993). The larvae are typical maggots, with a small anal segment with the spiracle endings.

Recognition

Xylota are elongate, fast-walking hoverflies, that look like sawflies (Hymenoptera, Symphyta). A group of species, including the common *X. segnis*, have a large red area on the abdomen. Others have a black abdomen with yellow to greyish spots (the abdomen is seldom fully black). Differences from *Chalcosyrphus* and *Brachypalpoides* are subtle. The key is based on Van der Goot (1981), Lucas (1981), Zeegers (1988), Beuk (1988), Verlinden (1991), Barendregt (1991), Speight (1999), Bartsch et al. (2002) and Doczkal (2004).

Key

1.a. Legs partly pale (whitish or yellowish) **>** 2

1.b. Legs entirely black; tergites black, with yellow spots on tergites 2 and 3. 10 mm. Northern Europe, Northern Asia **>** *Xylota suecica* Ringdahl
Note: if abdominal dorsum black with tergites 2 and 3 mostly red: *Brachypalpoides lentus*.

2.a. Baso-ventral ridge (if present) on hind tibiae bare; femur 3: ventral side without long bristles, but may be covered with scattered short black bristles **>** 3

2.b. Baso-ventral ridge on hind tibiae covered with a row of short, black spines (figure 791); femur 3 ventrally with a row of widely separated, long bristles on its top 1/2, in addition to scattered black bristles (figure 792); habitus figure 793.

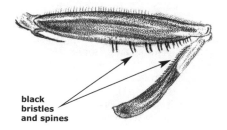

black bristles and spines

figure 791. *Xylota segnis*, leg 3 (Verlinden).

figure 792. *Xylota segnis*, femur 3 from below (Verlinden).

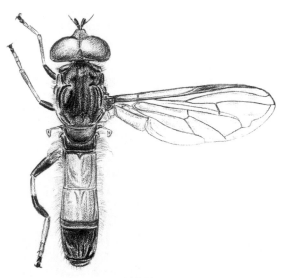

figure 793. *Xylota segnis*, habitus of male (Verlinden).

Holarctic species. 11-13 mm. Europe, in Asia into Japan, Nearctic **>** *Xylota segnis* Linnaeus
Jizz: relatively small *Xylota*, thorax greenish, abdomen black with large red patch.

3. Abdominal tergite 4 entirely or almost entirely covered with golden or whitish-yellow hairs (some short black hairs may be present along the basal margin, especially medially, but only within the basal 1/2 of the surface of the tergite), figure 794 ❯ 4
3.b. Abdominal tergite 4 black-haired over most of its surface and entirely black-haired medially from base to apex ❯ 6

4.a. Abdominal tergite 2 longer than wide or only slightly (less than 1 1/4 times) wider than long; adpressed hairs on abdominal tergites brightly golden. Male hind trochanter with 2 blunt spikes ❯ 5
4.b. Abdominal tergite 2 more than 1 1/2 times as wide as long; adpressed abdominal hairs only vaguely golden, more a faded whitish-yellow colour. Male: hind trochanter with 1 blunt spike. 12-14 mm. Northern Europe and mountains of Central Europe, in Asia to Mongolia ❯ *Xylota triangularis* Zetterstedt

5.a. Tibiae black on apical 1/3; habitus figure 794. Larger: 12-14 mm. Europe, Turkey ❯ *Xylota sylvarum* Linnaeus
Jizz: large *Xylota*, thorax blackish, abdomen black with golden patches.
5.b. Tibia entirely yellow. Smaller: 11-12 mm. Central Europe, in Asia into the Caucasus ❯ *Xylota xanthocnema* Collin
Jizz: medium-sized *Xylota*, thorax blackish, abdomen black with golden patches.

6.a. Hind tibiae yellow only at the base (as figure 791); hind metatars dark brown/black (except in *X. triangularis* female) ❯ 7
6.b. Hind tibiae widely yellow at both ends (figure 795); hind metatars (and 2 succeeding tarsal segments) yellow; abdominal tergites 2 and 3 with orange bands. Male: hypopygium with mixed white and black hairs. Palaearctic except the Mediterranean. 11-13 mm. Europe, in Asia into Japan ❯ *Xylota ignava* Panzer
Jizz: thorax bluish, abdomen black with large red patch.

7.a. Males (eyes meeting above the antennae) ❯ 8
7.b. Females (eyes not meeting above the antennae) ❯ 14

8.a. At least tergite 3 with yellowish or reddish markings ❯ 9
8.b. Tergites 2 and 3 entirely without pale markings, varying from slightly longer than broad to slightly broader than long; metatars 1 with 1 or 2 long, white, bristly hairs dorsoapically. 10-12 mm. Central and Northern Europe ❯ *Xylota caeruleiventris* Zetterstedt

9.a. Hypopygium white-haired, at most mixed with some black hairs ❯ 10
9.b. Hypopygium black-haired, at most mixed with some white hairs ❯ 13

10.a. Abdominal tergite 2 longer than wide ❯ 11
10.b. Abdominal tergite 2 wider than long ❯ 12

figure 794. *Xylota sylvarum*, habitus of male (Verlinden).

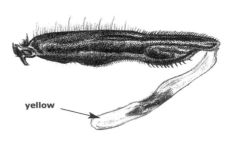

yellow

figure 795. *Xylota ignava*, leg 3 of male (Verlinden).

11.a. Abdomen: black with large red patch, which may be darkened (figure 796, figure 797); metatars 1 with a long, white, bristly hair dorso-apically, which reaches to the apical margin of the 2nd tarsal segment; pale hairs on the basal 1/2 of the antero-dorsal surface of femur 3 are of uniform length, none of them longer than 1/4 the maximum depth of femur; wing membrane not infuscated. Palaearctic except the Mediterranean. 9-11 mm. Europe, in Asia to Pacific coast **>** *Xylota tarda* Meigen

Jizz: abdomen black with red patch, very similar to *X. segnis*.

11.b. Abdomen: black with yellow spots, which may be darkened (figure 798); metatars 1 without a long, white, bristly hair on the dorsal surface; pale hairs on the basal 1/2 of the antero-dorsal sur-face of femur 3 of uneven length, some of them as long as 1/3 the maximum depth of the femur (figure 799); wings brownish over much of the apical 1/2 of the surface. 9-11 mm. Central Europe, in Asia into Japan. **>** *Xylota meigeniana* Stackelberg

Jizz: wings infuscated, abdomen black with yellow spots, tibia 3 pale on basal 1/3; anterior anepister-num shiny.

figure 797. *Xylota tarda*, abdomen of female (Verlinden).

figure 796. *Xylota tarda*, abdomen of male (Verlinden).

figure 798. *Xylota meigeniana*, habitus of male (Verlinden).

799.

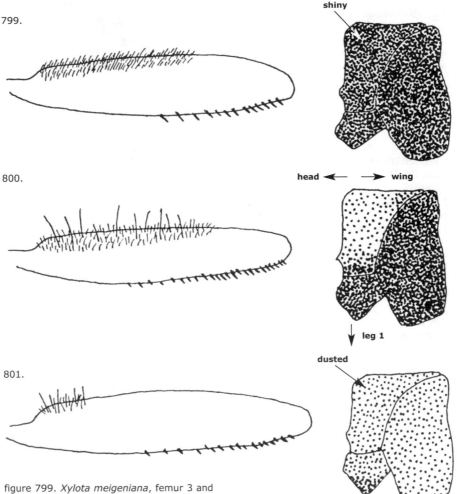

800.

shiny

head ◀── ──▶ wing

leg 1

dusted

801.

figure 799. *Xylota meigeniana*, femur 3 and anepisternum.
figure 800. *Xylota florum*, femur 3 and anepisternum.
figure 801. *Xylota jakutorum*, femur 3 and anepisternum (Beuk).

12.a. Thoracic dorsum with a transverse band of black hairs between the wing bases. Larger: 12-14 mm. Northern Europe and mountains of Central Europe, in Asia to Mongolia ❯ *Xylota triangularis* Zetterstedt
12.b. Thoracic dorsum completely pale-haired. Smaller: 8-10 mm. Central Europe, in Asia to Pacific coast. ❯ *Xylota abiens* Meigen
Jizz: small *Xylota*, abdomen black with yellow squarish spots, tibia 3 only pale at very base.

13.a. Hairs on the antero–dorsal surface of the hind femora including many at least as long as 1/2 the maximum depth of the hind femur, these longer hairs present on more than 1/2 the femur length; anterior anepisternum dull, posterior anepister-num shiny (figure 800). 11-13 mm. Northern and Central Europe, Western Siberia ❯ *Xylota florum* Fabricius
Jizz: large *Xylota*, abdomen black with elongate yellow spots, tibia 3 pale on basal 1/3.
13.b. Hairs on the antero–dorsal surface of the hind femora all shorter than 1/2 the maximum depth of a hind femur, longer hairs confined to the basal 1/5 of the femur length; posterior anepister-num dusted (figure 801). 10-12 mm. Northern and Central Europe, Siberia. ❯ *Xylota jakutorum* Bagatshanova

14.a. Anterior anepisternite 1 with most of surface undusted, brightly shiny (figure 799) ❯ 15
14.b. Anterior anepisternite with either entire surface, or most of surface dull, dusted (figure 800) ❯ 16

15.a. Abdominal tergite 3 with a transverse, orange band across anterior 1/2 of the tergite (figure 797); hind femora with middle 1/3 of ventral surface covered in black, spiny hairs. 9-11 mm. Europe, in Asia to Pacific coast ❯ *Xylota tarda* Meigen
15.b. Abdominal tergite 3 with a pair of pinkish markings, which may be reduced or obscure (as in figure 798); hind femora with middle 1/3 of ventral surface almost entirely covered in adpressed yellow spiny hairs, any black spiny hairs intermixed being mostly along the lateral margins; frons with triangular dust spots (figure 802). 9-11 mm. Central Europe, in Asia into Japan. ❯ *Xylota meigeniana* Stackelberg

16.a. Tars 3 with metatars and 2nd segment partly or mostly brownish-yellow dorsally (always pale apically), contrasting sharply in colour with the more distal black segments (pale hairs on abdominal tergites 2 and 3 yellow or whitish). 12-14 mm. Northern Europe and mountains of Central Europe, in Asia to Mongolia. ❯ *Xylota triangularis* Zetterstedt
16.b. Tars 3 almost entirely black dorsally (pale hairs on abdominal tergites 2 and 3 whitish) ❯ 17

17.a. Longest hairs on the antero-dorsal surface of the femur 3 no more than 1/4 as long as the maximum depth of the femur; posterior anepisternum dusted, dull (figure 801) ❯ 18
17.b. Longest hairs on antero-dorsal surface of femur 3 noticeably more than 1/3 as long as the maximum depth of the femur (nearly 1/2 the depth of the femur); posterior anepisternum shiny (figure 800). 11-13 mm. Northern and Central Europe, Western Siberia ❯ *Xylota florum* Fabricius

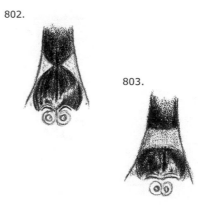

802.

803.

figure 802. *Xylota meigeniana*, frons of female.
figure 803. *Xylota abiens*, frons of female (Verlinden).

18.a. Metatars 1 with black bristles ventro-laterally; tibia 3 pale yellow on basal 1/3 of length; frons: band of dust just interrupted in the middle ❯ 19
18.b. Metatars 1 without black bristles; tibia 3 pale yellow on at most basal 1/5 of length; frons with entire band of dust. 8-10 mm. Central Europe, in Asia to Pacific coast ❯ *Xylota abiens* Meigen

19.a. Femur 2: ventral part of tip largely covered with microtrichia on hairless area, appearing dusted; tergites 2 and 3 without pale markings; thoracic dorsum pale-haired above wing base mixed with no, few, but occasionally many black hairs. 10-12 mm. Northern Europe ❯ *Xylota caeruleiventris* Zetterstedt
19.b. Femur 2: ventral part of tip largely or entirely bare of microtrichia on hairless area; tergites 2 and 3 with pale markings, which may be absent; thoracic dorsum with patch of black bristly hairs above wing base, with many more than 10 black hairs. 10-12 mm. Northern and Central Europe, Siberia ❯ *Xylota jakutorum* Bagatshanova
Note: females of these species remain difficult to identify because characteristics have some overlap.

INDEX

Species Index

	page number
Cheilosia carbonaria Egger 1860	67, 72
Cheilosia chloris (Meigen) 1822	79
Cheilosia chrysocoma (Meigen) 1822	77
Cheilosia cynocephala Loew 1840	73
Cheilosia fasciata Schiner & Egger 1853	66
Cheilosia flavipes (Panzer) 1798	60,79
Cheilosia fraterna (Meigen) 1830	79
Cheilosia frontalis Loew 1857	63
Cheilosia gigantea (Zetterstedt) 1838	68, 71
Cheilosia griseiventris Loew 1857	64
Cheilosia grossa (Fallén) 1817	76
Cheilosia himantopa (Panzer) 1798	78
Cheilosia illustrata (Harris) 1776	63
Cheilosia impressa Loew 1840	66
Cheilosia ingerae Nielsen & Claussen 2001	69, 70
Cheilosia lasiopa Kowartz 1885	62, 63
Cheilosia laticornis Rondani 1857	61
Cheilosia latifrons (Zetterstedt) 1843	56, 64
Cheilosia lenis Becker 1894	71, 73
Cheilosia longula (Zetterstedt) 1838	59
Cheilosia melanopa (Zetterstedt) 1843	62, 64
Cheilosia melanura Becker 1894	79
Cheilosia morio (Zetterstedt) 1838	63, 67
Cheilosia mutabilis (Fallén) 1817	59, 72
Cheilosia nebulosa Verrall 1871	77
Cheilosia nigripes (Meigen) 1822	58
Cheilosia orthotricha Vujic & Claussen 1994	78
Cheilosia pagana (Meigen) 1822	61
Cheilosia pallipes (Loew) 1863	59
Cheilosia personata Loew 1857	57
Cheilosia proxima (Zetterstedt) 1843	69, 71
Cheilosia psilophthalma Becker 1894	73
Cheilosia pubera (Zetterstedt) 1838	58
Cheilosia ranunculi Doczkal 2000	57, 65
Cheilosia rotundiventris Becker 1894	75
Cheilosia ruficollis Becker 1894	76
Cheilosia rufimana Becker 1894	68, 70
Cheilosia sahlbergi Becker 1894	58
Cheilosia scutellata (Fallén) 1817	60
Cheilosia semifasciata Becker 1894	66
Cheilosia sootryeni Nielsen 1970	75
Cheilosia soror (Zetterstedt) 1843	59
Cheilosia subpictipennis Claussen 1998	77
Cheilosia urbana (Meigen) 1822	72
Cheilosia uviformis Becker 1894	61, 68, 73
Cheilosia variabilis (Panzer) 1798	62
Cheilosia velutina Loew 1840	68, 69
Cheilosia vernalis (Fallén) 1817	75
Cheilosia vicina (Zetterstedt) 1849	57
Cheilosia vulpina (Meigen) 1822	64
Chrysogaster basalis Loew 1857	81
Chrysogaster coemiteriorum (Linnaeus) 1758	81
Chrysogaster rondanii Maibach & Goeldlin 1995	82
Chrysogaster solstitialis (Fallén) 1817	81
Chrysogaster virescens Loew 1854	82
Chrysosyrphus nasutus (Zetterstedt) 1838	82
Chrysosyrphus niger (Zetterstedt) 1843	82
Chrysotoxum arcuatum (Linnaeus) 1758	84
Chrysotoxum bicinctum (Linnaeus) 1758	84

Species Index

	page number

LITERATURE

Barendregt A. 1978. Zweefvliegentabel, zesde druk. Jeugdbondsuitgeverij, Utrecht pp: 1-82.

Barendregt A. 1982. Zweefvliegentabel, zevende druk. Jeugdbondsuitgeverij, Utrecht, pp: 1-82.

Barendregt A. 1983. *Syrphus nitidifrons*, Becker, 1921, from the Netherlands, with description of the male, and a key to the European *Syrphus* species (Diptera: Syrphidae). Entomologische Berichten Amsterdam 43: 59-64.

Barendregt A. 1991. Zweefvliegentabel, achtste druk. Jeugdbondsuitgeverij, Utrecht pp: 1-92.

Barendregt A. Van Steenis W. and Van Steenis J.. 2000. *Spilomyia* species (Diptera: Syrphidae) in Dutch collections, with notes on their European distribution. Entomologische Berichten Amsterdam 60: 41-45.

Barendregt A. 2001. Zweefvliegentabel, negende druk. Jeugdbondsuitgeverij, Utrecht, pp: 1-92.

Barkalov A.V. and Stahls G. 1997. Revision of the Palaearctic bare-eyed and black-legged species of the genus *Cheilosia* Meigen (Diptera, Syrphidae). Acta Zoologica Fennica 208: 1-74.

Barkemeyer W. 1986. Zum Vorkommen seltener und bemerkenswerter Schwebfliegen in Niedersachsen (Diptera, Syrphidae). Drosera 86: 79-88.

Barkemeyer W. and Claussen C. 1986. Zur Identitat van *Neoascia unifasciata* (Strobl, 1898) - mit einem Schlussel fur die in der Bundesrepublik Deutschland nachgewiesene Arten der Gattung *Neoascia* (Williston, 1886) (Diptera: Syrphidae). Bonner Zoologischer Beitrage 37: 229-239.

Barkemeyer W. 1994. Untersuchung zum Vorkommen der Schwebfliegen in Niedersachsen und Bremen (Diptera: Syrphidae). Naturschutz und Landschaftspflege in Niedersachsen 31: 1-514.

Barkemeyer W. 1997. Zur Ökologie der Schwebfliegen und andere Fliegen Urbaner Bereiche (Insecta: Diptera). Archiv Zoologischer Publicationen Band 3 M Galunder Verlag, Wiehl. Germany.

Bartsch H. 1997. Efterlysta, ovanliga, föbisedda och några andra intressanta noorländska blomflugor (Diptera, Syrphidae). Natur i Norr 16: 69-94.

Bartsch H.D. 2001. Swedish Province Catalogue for Hoverflies (Diptera, Syrphidae). Entomologisk Tidskrift 122: 189-215.

Bartsch H. and Bergström C. 2001. Blom- og parasitflugor från Lycksele lappmark, särskilt Ajaure (Diptera: Syrphidae, Tachinidae). Natur I Norr, Umeå 21.

Bartsch H.D., Nielsen T.R. and Speight M.C.D. 2002. Reappraisal of *Xylota caeruleiventris* Zetterstedt 1838 with remarks on the distribution of this species and *X. jacutorum* Bagatshanova 1980 in Europe. Volucella 6: 69-80.

Bartsch H., Binkiewicz E., Rådén A. and Nasibov E. 2009a. Nationalnyckeln till Sveriges flora och fauna 53a. Tvåvingar: Blomflugor: Syrphinae. Diptera: Syrphidae: Syrphinae. Artdatabanken, SLU, Uppsala.

Bartsch H., Binkiewicz E., Klintbjer A. Rådén A. and Nasibov E. 2009b. Nationalnyckeln till Sveriges flora och fauna 53b. Tvåvingar: Blomflugor: Eristalinae and Microdontinae. Diptera: Syrphidae: Eristalinae & Microdontinae. Artdatabanken, SLU, Uppsala.

Beuk P. 1988. Redactie: Xylota's. Stridula 12(2): 12 13.

Brådescu V. 1991. Les Syrphides de Roumanie (Diptera, Syrphidae), Clés de détermination et répartition. Travaux des. Museum d'istoire. Naturelle Grigore Antipa 31: 7-83.

Claussen C. and Torp E. 1980. Untersuchungen über vier europäischen Arten der Gattung *Anasimyia* Schiner, 1864 (Insecta, Diptera, Syrphidae) Mitteilungen Zoologischer Museum Universität Kiel 1:16.

Claussen C. 1984. *Sphaerophoria potentillae* n. sp. - eine neue Syrphiden-Art aus Nordwestdeutschland (Diptera: Syrphidae). Entomologische Zeitschrift 94: 245-250.

Claussen C. and Speight M.C.D. 1988. Zur Kenntnis von *Cheilosia vulpina* (Meigen, 1822) und *Cheilosia nebulosa* Verrall, 1871 (Diptera, Syrphidae). Bonner Zoologischer Beitrage. 39(1): 19-28.

Claussen C. and Kassebeer C.F. 1993. Eine neue Art der Gattung *Cheilosia* Meigen 1822 aus den Pyrenäen (Diptera: Syrphidae). Entomologische Zeitschrift 103: 420-427.

Claussen C. and Vujic A. 1995. Eine neue Art der Gattung *Cheilosia* Meigen aus Mitteleurope (Diptera: Syrphidae). Entomologische Zeitschrift 105: 77-85.

Claussen C. 1998. Die europaischen Arten der *Cheilosia alpina*-gruppe (Diptera, Syrphidae). Bonner Zoologischer Beitrage. 47: 381-410.

Claussen C. and Doczkal D. 1998. Ein neue Art der Gattung *Cheilosia* Meigen, 1822 (Diptera, Syrphidae) aus den Zentralalpen. Volucella 3: 1-13.

Claussen C. and Speight M.C.D. 1999. On the identity of *Cheilosia ruralis* (Meigen, 1822) (Diptera, Syrphidae) – with a review of its synonyms. Volucella 4: 93-102.

De Buck N. 1990. Bloembezoek en bestuivingsecologie van Zweefvliegen (Diptera, Syrphidae) in het bijzonder voor Belgie. Studiedocumenten van het K.B.I.N. 60. K.B.I.N. Brussels. pp: 1-167.

Derksen W. 1941. Die Succession der pterygoten Insekten im abgestorben Buchenholz. Zeitschrift. Morphologie und. Okologie der Tiere 37: 683-734.

Dirickx H.G. 1994. Atlas des Diptères syrphides de la région méditerranéenne. Studiedocumenten van het K.B.I.N., K.B.I.N., Brussels, pp: 1-317.

Doczkal D. and Schmid U. 1994. Drei neue Arten der Gattung *Epistrophe* (Diptera: Syrphidae), mit einem Bestimmungsschlussel for die deutchen Arten. Stuttgarter Beitrage zur Naturkunde serie A: 507: 1-32.

Doczkal D. 1995. Bestimmungsschlüssel für die Weibchen der deutschen *Sphegina*-arten (Diptera, Syrphidae). Volucella 1: 3-19.

Doczkal D. 1996. Schwebfliegen aus Deutschland: Erstnachweise und wenig bekante Arten (Diptera, Syrphidae). Volucella 2: 36-62.

Doczkal D. 1998. *Leucozona lucorum* (Linnaeus) - a species complex? Volucella 3: 27-50.

Dozckal D. and Schmid U. 1999. Revision der mitteleuropaischen Arten der Gattung *Microdon* Meigen (Diptera, Syrphidae). Volucella 4: 45-68.

Doczkal D. 2000a. Description of *Cheilosia ranunculi* spec. nov. from Europe, a sibling species of *C. albitarsis* Meigen (Diptera, Syrphidae). Volucella 5: 63-78.

Doczkal D. 2000b. Redescription of *Leucozone nigripila* Mik and description of *Leucozona inopinata* spec. nov. (Diptera, Syrphidae). Volucella 5: 115-127.

Doczkal D., Stuke J.-H. and Goeldin de Tiefeneau P. 2002. The species of the *Platycheirus scutatus* (Meigen) complex in central Europe, with description of *Platycheirus speighti* spec. nov. from the Alps (Diptera, Syrphidae). Volucella 6: 23-40.

Dozckal D. 2004. *Xylota caeruleiventris* Zetterstedt, 1838 (Diptera, Syrphidae) found in central Europa, with remarks on the identification of the female. Volucella 7: 193-200.

Doczkal D. and Dziock F. 2004. Two new species of *Brachyopa* Meigen from Germany, with notes on *B. grunewaldensis* Kassebeer (Diptera, Syrphidae). Volucella 7: 35-60.

Dusek J. and Laska P. 1961. Beitrag zur Kenntnis der Schwebfliegen-Larven III (Syrphidae, Diptera). Prirod. cas. slezsky 22: 513-541.

Dusek J. and Laska P. 1976. European species of *Metasyrphus*: key, descriptions and notes (Diptera: Syrphidae). Acta Entomological Bohemoslovaka 73: 263-282.

Dusek J. and Laska P. 1982. European species related to *Platycheirus manicatus*, with descriptions of two new species (Diptera, Syrphidae). Acta Entomologica Bohemoslavica 79: 377-392.

Dziok F. 2002. Überlebensstrategieen un Nahrungsspezialisierung bei räuberischen Schwebfliegen (Diptera, Syrphidae). Dissertation 10/2002 UFZ-Leipzig. pp: 1-131.

Gilbert F.S. 1993. Hoverflies. Naturalists Handbook 5, Richmond Publishing Co. Ltd., Slough,. pp: 1-67.

Goeldlin de Tiefeneau P. 1974. Contribution a l'etude systematique et ecologique des Syrphides (Diptera) de la Suisse occidentale. Mitteilungen der Schweizerischen Entomologischen Gesellschaft 47: 151-252.

Goeldlin de Tiefeneau P. 1976. Revision du genre *Paragus* (Dipt. Syrphidae) de la region palearctique occidentale. Mitteilungen der Schweizerischen Entomologischen Gesellschaft 49: 79-108.

Goeldlin de Tiefenau P. 1989. Sur plusieurs especes de *Sphaerophoria* (Dipt., Syrphidae) nouvelles ou meconnues de regions palearctiques et nearctique. Mitteilungen der Schweizerischen Entomologischen Gesellschaft 62: 41-66.

Goeldlin de Tiefenau P., Maibach A. and Speight M.C.D. 1990. Sur quelques especes de Platycheirus (Diptera: Syrphidae) nouvelle ou malconnues. Dipterists Digest 5: 19-44.

Goeldlin de Tiefenau P. 1991. *Sphaerophoria estebani*, une nouvelle espece europeenne de groupe rueppellii (Diptera, Syrphidae). Mitteilungen der Schweizerischen Entomologischen Gesellschaft 64: 331-339.

Goeldlin de Tiefenau P. 1996. Sur plusieurs nouvelles especes europeennes de *Syrphus* (Diptera, Syrphidae) et cle des especes palearctiques du genre. Mitteilungen der Schweizerischen Entomologischen Gesellschaft 69: 157-171.

Goeldlin de Tiefenau P. 1997. Le genre
Trichopsomyia Williston, 1888 (Diptera:
Syrphidae) en Europe avec description d'une
nouvelle espèce, connue depuis longtemps.
Bull. Soc. Ent. Suisse 70: 191-201.

Hartley J.C. 1961. A taxonomic account of the
larvae of some British Syrphidae. Proc.
Zool. Soc. Lond. 136: 505-573.

Hippa H., Nielsen T.R. and Van Steenis J. 2001.
The West Palaearctic species of the genus
Eristalis Latreille (Diptera, Syrphidae).
Norwegian Journal of Entomology 48:
289-327.

Holloway G.J. 1993. Phenotypic variation in
colour pattern and seasonal plasticity in
Eristalis hoverflies (Diptera: Syrphidae).
Ecological Entomology 18: 209-217.

Hurkmans W. 1993. A Monograph of Merodon
(Diptera: Syrphidae). Part 1. Tijdschrift voor
Entomologie 136: 147-234.

Kanervo E. 1938. Zur Systematik und
Phylogenie der westpaläarktischen Eristalis-
Arten (Dipt. Syrphidae) mit einer Revision
derjenigen Finnlands. Annales Universite
Turkuensis. Serie A 6(4): 1-54.

Kassebeer C.F. 1991. Eine neue Art der Gattung
Sphegina Meigen 1822 aus Europa
(Diptera: Syrphidae). Entomologische
Zeitschrift 101: 441-446.

Kassebeer C.F. 1995. Revision der paläarktis-
chen Arten der Gattung Chrysosyrphus
Sedman, 1965 (Diptera, Syrphidae). Studia
dipterologica 2: 283-295.

Kormann K. 1988. Schwebfliegen
Mitteleuropas: Vorkommen, Bestimmung,
Beschreibung. Ecomed, Munchen: 1-176.

Kormann K. 1993. Schwebfliegen aus der
Umgebung von Karlsruhe. Zeit. Ent. 14:
33-56.

Krivosheina N.P. 2004. Morphology of the
species of the genus Temnostoma from apiforme and vespiforme groups. Zoologitseskii
Zhurnal 82: 1475-1486.

Kula E. 1983. The larva and puparium of
Eriozona syrphiodes (Fallén) (Diptera,
Syrphidae). Acta ent. Bohemoslovaca 80:
71-73.

Lucas J.A.W. 1981. Syrphiden allerlei.
Entomologische Berichten Amsterdam 41:
49-53.

Lucas J, 1992. Een nieuwe zweefvliegsoort voor
Nederland: Scaeva dignota (Rondani, 1857).
Vliegenmepper 2: 1-2.

Maibach A. and Goeldlin de Tiefenau P. 1989.
Mallota cimbiciformis (Fallén) nouvelle pour
la faune de Suisse: morphologie du dernier
stade larvaire, de la pupe et notes biolo-
giques (Diptera, Syrphidae). Bulletin de.
Societe. Entomologique Suisse 62: 67-78.

Maibach A., Goeldlin de Tiefenau P. and
Speight M.C.D. 1994a. Limites generiques
et characteristiques taxonomiques de
plusieurs genres de la tribu des chrysogasterini (Diptera: Syrphidae). I Diagnoses
generiques et description de Riponnesia
gen. nov. Annales de. Societe Entomologique de France. (N.S.) 30: 217-247.

Maibach A., Goeldlin de Tiefenau P. and
Speight M.C.D. 1994b. Limites generiques
et characteristiques taxonomiques de
plusieurs genres de la tribu des chrysogasterini (Diptera: Syrphidae) II Statut tax-
onomique de plusieurs des especes
etudiees et analyse du complexe
Melanogaster macquarti (Loew). Annales de.
Societe Entomologique de France. (N.S.)
30: 253-271.

Maibach, A. and Goeldlin de Tiefenau, P. 1994.
Limites génériques et charactéristiques taxonomiques de plusieurs genres de la Tribu
des Chrysogasterini (Diptera: Syrphidae)
III. Descriptions des stades immatures de
plusieurs espèces ouest- paléarctiques. Rev.
Suisse Zool. 101: 369-411.

Maibach A. and Goeldlin de Tiefenau P. 1995.
Chrysogaster rondanii sp. n. from Western
and Central Europe (Diptera: Syrphidae).
Mitteilungen Schweizerischen
Entomologischen Geselschaft 68: 459-464.

Mazanek L., Laska P. and Bicik V. 1999a.
Revision of type material of Metasyrphus
chillcotti (Diptera: Syrphidae). Dipterologica
Bohemoslovaca 9: 139-142.

Mazanek L., Laska P., Bicik V. and Nielsen T.R.
1999b. Key to males of Norwegian species
of Eupeodes (Diptera: Syrphidae).
Dipterologica Bohemoslovaca 9: 143-152.

Mazanek L., Laska P., and Bicik V. 1999c. Two
new Palaearctic species of Eupeodes similar
to E. bucculatus (Diptera, Syrphidae).
Volucella 4: 1-9.

Mazanek L., Laska P., Bicik V. and Nielsen T.R.
2004. Key to females of Norwegian species
of Eupeodes (Diptera, Syrphidae).
Dipterologica Bohemoslovaca 14.

Nielsen T. R. 1970. Cheilosia sootryeni nov. sp.
(Dipt., Syrphidae), a Norwegian species
resembling Ch. vernalis Fallén. Norsk Ent.
Tidssk. 17(2): 115-118.

Nielsen T.R. 1981. Studies on Platycheirus
Lepeletier & Serville: P. complicatus Becker,
P. latimanus (Wahlberg) and P. boreomontanus
nom. nov. (Diptera: Syrphidae)
Entomologica Scandinavica 12: 99-102.

Nielsen T.R. 1995. Studies on some northern
Eristalis species (Diptera, Syrphidae).
International Journal of Dipterological
Research 6: 129-133.

Nielsen T.R. 1997. The hoverfly genera *Anasimyia* Schiner, *Helophilus* Meigen, *Parhelophilus* Girschner and *Sericomyia* Meigen in Norway. Fauna Norvegica Serie B 44: 107-122.

Nielsen T.R. 1999. Check-list and distribution maps of Norwegian Hoverflies, with description of *Platycheirus laskai* nov. sp. (Diptera, Syrphidae). NINA Fagrapport 035: 1-99.

Nielsen T.R. and Claussen C. 2001. On *Cheilosia ingerae* spec. nov. (Diptera, Syrphidae) from northern Fennoscandia. Dipteron 4: 43-56.

Nielsen T.R. 2003. Descrption of *Eupeodes biciki* spec. nov. (Diptera, Syrphidae) from northern Norway. Norwegian Journal of Entomology 50: 99-103.

Nielsen T.R. 2004. European species of the *Platycheirus ambiguus* group (Diptera, Syrphidae), with description of new species. Volucella 7: 1-30.

Peck L.V. 1988. Syrphidae, pp: 11- 230 in: Soos and Paap, Catalogue of Palaearctic Diptera, Syrphidae-Conopidae, Volume 8.

Pettersson R.B. and Bartsch H. D. 2001. *Blera eoa* (Stackelberg, 1928), en ny stubb-blomfluga för Europa (Diptera, Syrphidae). Natur I Norr, Umeå 20: 91-96.

Reemer M. 2000a. A new species of *Parhelophilus* Girschner, 1897 (Diptera, Syrphidae) from southwest Europe. Dipteron 3: 1-6.

Reemer M. 2000b. Zweefvliegenveldgids (Diptera, Syrphidae). Jeugdbondsuitgeverij, Utrecht.

Reemer M. 2002. Lena's wimperzweefvlieg *Dasysyrphus lenensis* in Nederland (Diptera: Syrphidae). Nederlandse Faunistische Mededelingen 17: 13-18.

Renema W. and Wakkie B. 2001. Het zweefvlieggenus *Callicera* in Nederland en België (Diptera: Syrphidae). Nederlandse Faunistische Mededelingen 14: 1-12.

Röder G. 1990. Biologie der Schwebfliegen Deutschlands (Diptera: Syrphidae). Erna Bauer Verlag, Keltern-Weiler: 1-136.

Rotheray G.E. 1990. Larval and puparial records of some hoverflies associated with dead wood (Diptera, Syrphidae). Dipterists Digest 7: 2-7.

Rotheray G.E. 1993. Colour guide to hoverfly larvae. Dipterists Digest 9,: 1-156.

Rotheray G.E. 1996. The larva of *Brachyopa scutellaris* Robineau-Desvoidy (Diptera: Syrphidae) with a key to and notes on the larvae of British Brachyopa species. Entomologist's Gazette 47: 199-205.

Rotheray G.E. 1998. *Platycheirus splendidus* sp. n. from Britain formerly confused with *Platycheirus scutatus* (Diptera: Syrphidae). Entomologist's Gazette 49: 271-276.

Rotheray G.E. and Stuke J.H. 1998. Third stage larvae of four species of saproxylic Syrphidae (Diptera), with a key to the larvae of British *Criorhina* species. Entomologist's Gazette 49: 209-217.

Sack P. 1930. Schwebfliegen oder Syrphidae, pp: 2-118 in Dahl F. (ed.) Zweiflügler oder Diptera IV: Syrphidae – Conopidae. Die Tierwelt Deutchlands und der angrenzenden Meeresteile.

Sack P. 1932. Syrphidae, in: Lindner E. Fliegen der Palaearktische Region, Teil 31. E. Nagele, Stuttgart.

Schmid U. 1996. Auf glasern Schwingen: Schwebfliegen. Stuttgarter beitrage zur Naturkunde, serie C, heft 40: 1-81.

Schmid U. 1999. *Syrphus obscuripes* Strobl, 1910: ein älteres Synonym von *Epistrophe similis* Doczkal & Schmid, 1994 (Diptera, Syrphidae). Volucella 4: 103-104.

Schönrogge K., Barr B., Wardlaw J.C., Napper E., Gardner M.G., Breen J., Elmes G.W. and Thomas J.A. 2002b. Addendum: When rare species become endangered: cryptic speciation in myrmicophilous hoverflies. Biological Journal of the Linnaean Society 76: 315.

Seguy E. 1961. Dipteres Syrphides de l'Europe Occidentale. Memoires du Museum National d'Histoire Naturelle. Nouvelle Serie, Serie A, Zoologie, Tome XXIII. Paris.

Smit J.T. 2001. De zweefvlieg *Platycheirus splendidus* nieuw voor Nederland, Belgie en Frankrijk (Diptera: Syrphidae). Nederlandse Faunistische Mededelingen 15: 141-148.

Smit J.T., Renema W., Van Aartsen B. 2001a. De zweefvliegen *Chrysotoxum intermedium* en *Chrysotoxum fasciolatum* nieuw voor de Nederlandse fauna (Diptera: Syrphidae). Nederlandse Faunistische Mededelingen 15: 117-122.

Smit J.T., Reemer M. and Renema W. 2001b. Vijf soorten van het genus *Cheilosia* nieuw voor Nederland (Diptera: Syrphidae). Nederlandse Faunistische Mededelingen 15: 123-140.

Smit J.T. 2002. Het zweefvlieg genus *Sphegina* in Nederland (Diptera: Syrphidae). Nederlandse Faunistische Mededelingen 16: 85-90.

Smit J.T. 2003. De zweefvlieg *Platycheirus auro-lateralis*, een tweede zustersoort van *P. scuta-tus*, nieuw voor Nederland (Diptera: Syrphidae). Nederlandse Faunistische Mededelingen 19: 87-94.

Smit J.T. and Zeegers T. 2005. overzicht van het zweefvliegengenus psilota in nederland (diptera: syrphidae). Nederlandse Faunistische Mededelingen 22: 113-120.Schönrogge K., Barr B.,Wardlaw J.C., Napper E., Gardner M.G., Breen J., Elmes G.W. and Thomas J.A. 2002a.When rare species become endangered: cryptic specia-tion in myrmicophilous hoverflies. Biological Journal of the Linnaean Society 75: 291-300.

Sommaggio D. 2002. *Paragus gorgus* Vujic & Radenkovic 1999: a junior synonym of *P. majoranae* Rondani, 1857, and reinstate-ment of *P. pecchiolii* Rondani, 1857 (Diptera, Syrphidae).Volucella 6: 53-56.

Sorokina V.S. 2002. Beschreibung van drei neue arten der gattung *Paragus* Latreille 1804 (Diptera, Syrphidae) aus Asien, mit einem bestimmungsschlüssel der bisher bekannten russischen *Paragus*-arten.Volucella 6: 1-22.

Speight M.C.D., De Courcy Williams M. and Legrand J. 1986. *Scaeva dignota* et *S. mecogramma* nouveaux pour la France et cle de determination des especes du genre (Diptera, Syrphidae). L'entomologiste 42: 359-364.

Speight M.C.D. 1988. Syrphidae known from temperate Western Europe: potential addi-tions to the fauna of Great Britain and Ireland and a provisional species list for N. France. Dipterists Digest 1: 2-35.

Speight M.C.D. and Goeldlin de Tiefenau P. 1990. Keys to distinguish *Platycheirus angustipes*, *P. europeus*, *P. occultus* and *P. ram-sarensis* (Dipt. Syrphidae) from other *clypea-tus* group species known in Europe. Dipterists Digest 5: 5-18.

Speight M.C.D. 1991a. A key to W European *Parasyrphus* species (Syrphidae). Dipterists Digest 8: 3-5.

Speight M.C.D. 1991b. *Callicera aenea, C. aurata, C. fagesii* and *C. macquartii* redefined, with a key to and notes on the European *Callicera* species (Diptera: Syrphidae). Dipterists Digest, 10: 1-25.

Speight M.C.D. 1999. A key to European *Xylotini* (Dip.: Syrphidae). Entomologist's Record 111: 211-218.

Speight M.C.D. 2002. Two controversial addi-tions to the Irish insect list: *Microdon myr-micae* Schönrogge et al. and *Pipiza festiva* Meigen (Diptera: Syrphidae). Bulletin of the Irish Biogeographical Society 26: 143-153.

Speight M.C.D. 2003. Species accounts of European Syrphidae (Diptera): species of the Atlantic, Continental and Northern Regions. pp: 1-209 in: Speight, M.C.D., Castella, E., Obrdlik, P. and Ball, S. (eds.) Syrph the Net, the database of European Syrphidae, vol. 39, Syrph the Net publica-tions, Dublin.

Ssymank A., Barkemeyer W., Claussen C., Löhr P-W. and Scholz A. 1999. Syrphidae, pp:195-203 in: Schumann H., Bährman R. and Stark A. Checkliste der Dipteren Deutschlands. EntomoFauna Germanica, Studia Dipterologica Supplement 2.

Ssymank A. 2001.Vegetation und blütenbe-suchende Insekten in der Kulturlandschaft. Schriftenreihe für Landschaftsplege und Naturschutz, Heft 64. Bundesambt für Naturschutz, Bonn.

Stackelberg A.A. 1970. Syrphidae, pp: 11-96 in: Bei-Bienko G.Ya. (ed.) Identification Keys to the Insects of the European part of the USSR (103), part 2. Nauka Publishers, Leningrad (In Russian).

Stackelberg A.A. 1988. 49. Family Syrphidae, pp: 10-148 in Bei-Bienko G.Ya. (ed.) Identification Keys to the Insects of the European part of the USSR (103), part 2. E.J. Brill Publishing Company, Leiden (English Translation of Stackelberg, 1970).

Stubbs A.E. and Falk S.J. 1983. British Hoverflies, an illustrated identification guide. British Entomological and Natural History Society, Reading.

Stubbs A.E. 1996. British Hoverflies, second supplement. British Entomological and Natural History Society, Reading.

Stubbs A.E. 2002. Advances in the understand-ing of the *Platycheirus scutatus* (Meigen) complex in Britain, including the addition of *Platycheirus aurolateralis* sp. n. (Diptera, Syrphidae). Dipterists Digest 9: 75-80.

Stubbs A.E. and Falk S.J. 2002. British Hoverflies, an illustrated identification guide. British Entomological and Natural History Society, Reading.

Stuke J.H. and Claussen C. 2000. *Cheilosia canic-ularis* auctt. - ein Artencomplex.Volucella 5: 79-94.

Thompson F.C. 1980. The problem of old names as illustrated by *Brachyopa "conica* Panzer", with a synopsis of Palaearctic *Brachyopa* Meigen (Diptera: Syrphidae). Entomologica Scandinavica 11: 209-216.

Thompson F.C. and Torp E. 1982. Two new Palaearctic Syrphidae (Diptera). Entomologica Scandinavica 13: 441-444.

Thompson F.C. and Torp E. 1986. Synopsis of the European Species of *Sphegina* Meigen (Diptera: Syrphidae). Entomologica Scandinavica 17: 235-269.

Thompson F.C. and Rotheray G. 1998. 3.5 Family Syrphidae: 81-139 in: Papp L. and Darvas B. (eds.) Contributions to a Manual of Palaearctic Diptera (with special reference to flies economic importance) Volume 3, Higher Brachycera.

Torp E. 1994. Danmarks Svirrefluer (Diptera: Syrphidae). Danmarks Dyreliv, Bind 6: 1-490. Apollo Books, Stenstrup.

Van der Goot V.S. 1975. Zweefvliegentabel, vijfde druk, CJN, KJN, NJN.

Van der Goot V. S. 1981. De zweefvliegen van Noordwest-Europa en Europees Rusland, in het bijzonder de Benelux. –Bibliotheek van de. K.N. N.V. 32: 1-274.

Van der Goot V.S. 1986. Aanvulling op het boek: De Zweefvliegen van Noordwest-Europa en Europees Rusland, in het bijzonder de Benelux, uitgave 32a, Bibliotheek van de KNNV. Hoogwoud.

Van der Goot V.S. 1986. Zweefvliegen in kleur. Bibliotheek van de KNNV no 32a, Hoogwoud.

Van der Goot V. 1989. Zweefvliegen. KNNV uitgeverij en Jeugdbondsuitgeverij, Utrecht.

Van der Linden J. 1986. Het voorkomen van het genus *Platycheirus* (Diptera: Syrphidae) in Nederland. Nieuwsbrief European Invertebrate Survey-Nederland 17: 3-22.

Van der Linden J. 1991. Nieuwe soorten van het genus *Platycheirus* in Nederland en Belgie (Diptera: Syrphidae). Entomologische Berichten Amsterdam 51: 112-116.

Van Steenis J. 1998a. *Rhingia borealis* nieuw voor Nederland en Belgie, met een tabel tot de Europese *Rhingia*-soorten (Diptera: Syrphidae). Entomologische Berichten Amsterdam 59: 73-77.

Van Steenis J. 1998b. Some rare hoverflies in Sweden (Diptera: Syrphidae). Entomologisk Tidskrift 119: 83-88.

Van Steenis J. and Goeldlin de Tiefenau P. 1998. Description of and key to the European females of the *Platycheirus peltatus* subgroup (Diptera, Syrphidae), with a description of the male and female of *P. islandicus* Ringdahl, 1930, stat. n. Mitteilungen der Schweizerischen Entomologischen Gesellschaft. 71: 187-199.

Van Steenis J. 2000. The West-Palaearctic species of *Spilomyia* Meigen (Diptera, Syrphidae). Mitteilungen der Schweizerischen Entomologische Gesellschaft 73: 143-168.

Van Steenis W. 1996. *Eristalis alpina* versus de *E. rupium*-groep. Vliegenmepper 5(2): 9.

Van Steenis W. and Barendregt A. 2002. Family Syrphidae, pp:200-216 in: Beuk P. Checklist of the Diptera of the Netherlands, KNNV Uitgeverij, Utrecht.

Van Veen M.P. and Zeegers Th. 1996. *Eristalis picea, Eristalis rupium* en *Eristalis vitripennis* revisited. Vliegenmepper 5(2): 7-8.

Van Veen M.P. and Zeegers Th. 1998. Faunistiek van *Eristalis picea* (Diptera: Syrphidae) in Nederland. Entomologische Berichten Amsterdam 58: 37-40.

Van Veen M.P. 1998. Zweefvliegen in een achtertuin. Vliegenmepper 7(2): 8-9.

Verlinden L. 1991. Zweefvliegen (Syrphidae). Fauna van België. Koninklijk Belgisch Instituut voor Natuurwetenschappen 39: 1-298, Brussels.

Verlinden L. 1994. Syrphides (Syrphidae). Faune de Belgique. Koninklijk Belgisch Instituut voor Natuurwetenschappen 39: 1-289, Brussels.

Verlinden L. 1999. A new *Pipizella* (Siptera, Syrphidae) from the French and Italian Alps, with a key to the *Pipizella* species of Central and Western Europe. Volucella 4: 11-28.

Verlinden L. 1999. *Cheilosia hypena* Becker, 1894 (Diptera, Syrphidae) – description of the male, redescription of the female and its separation from *Cheilosia frontalis* Loew, 1857. Volucella 4: 85-92.

Verlinden L. 2000. Some notes on the variability of *Cheilosia insignis* Loew, 1857 (Diptera, Syrphidae) with suggestions for adapting existing keys. Volucella 5: 103-114.

Violovitsh N.A. 1983. Siberian Syrphidae (Diptera). Translated by Van der Goot V.S. and Verlinden L. Verslagen en technische gegevens 43, ITZ, Univ. Amsterdam, Amsterdam.

Vockeroth J.R. 1992. The flower flies of the subfamily Syrphinae of Canada, Alaska and Greenland, Diptera: Syrphidae. The insects and arachnids of Canada Part 18. Agriculture Canada.

Vujic A. 1987. New species of genus *Sphegina* Meigen 1822 (Diptera, Syrphidae). Bull. Nat. Hist. Mus., Belgrade, B42: 79-83.

Vujic A. 1990. Genera *Neoascia* Williston 1886 and *Sphegina* Meigen 1822 (Diptera: Syrphidae) in Yugoslavia and description of species *Sphegina sublatifrons* sp. nova. Bulletin of the Natural History Museum, Belgrade, B45: 78-93.

Vujic A. and Claussen C. 1994a. *Cheilosia orthotricha* spec. nov., eine weitere Art aus der Verwandtschaft von *Cheilosia canicularis* aus Mitteleuropa (insecta, Diptera, Syrphidae). Spixiana 17: 261-267.

Vujic A. and Claussen C. 1994b. *Cheilosia bracusi*, a new hoverfly from the mountains of Central and Southern Europe (Diptera: Syrphidae). Bonner Zoologischer. Beitrag 45: 137-146.

Vujic A. and Stuke J.H. 1998. A new hoverfly of the genus *Melanogaster* from Central Europe (Diptera, Syrphidae). Studia Dipterologica 5: 343-347.

Vujic A., Radenkovic S. and Stănescu C. 1998. New data on hoverflies (Diptera, Syrphidae) in Romania. Volucella 3: 63-74.

Wakkie B. 2001. Zweefvlieglarven-excursie 18 februari 2001. Zweefvliegennieuwsbrief 5: 5-7.

Wolff D. 1998. *Pipiza accola* Violovitsh, 1985 (Diptera, Syrphidae) - Erstnachweis für Deutschland. Drosera 1998: 123-126.

Zeegers Th. 1988. Determinatieproblemen bij zweefvliegen II. Stridula 12(2): 3-11.

Zeegers Th., Van Veen M.P. and De Boer E.P. 1989. *Psilota anthracina* nieuw voor Nederland (Diptera: Syrphidae). Entomologische Berichten Amsterdam 49: 109-110.

Zeegers Th. and van Veen M.P. 1992. Over de zweefvliegen *Eristalis rupium*, *E. picea* en *E. fennica* in Nederland en België. Vliegenmepper 2: 10-14.

Zeegers T. 2000. Determinatie van *Cheilosia's* van de *bergenstammi*-groep. Zweefvliegennieuwsbrief 4: 11-16.

Note: Speight (2003) gives an almost complete list of European hoverfly literature.

COLOPHON

Author: Mark P. van Veen, The Netherlands
Edited by: Suzanne J. Moore, New Zealand
Technical advice: Ad Littel and Nico Schonewille,
The Netherlands, and Martin Speight, Ireland
Design: Varwig Design, Hengelo, The Netherlands

This publication was supported financially by:
The Uytenboogaart-Eliasen foundation
Foundation Funds KNNV

Cover photography: M.P. van Veen
Front: large photo, *Episyrphus balteatus* (female, Zeist);
small photos from left to right: *Cheilosia grossa* (male,
Zeist); *Eumerus tricolor* (male, Wrakelberg); *Helophilus
trivittatus* (male, Plateaux); Myathropa florea (male,
Plateaux); *Melanogaster nuda* (male, Naardermeer)
Back: from left to right, *Merodon equestris* (female, Zeist);
Eristalis lineata (male, Naardermeer); *Cheilosia variabilis*
(female, Zeist).

© KNNV Publishing, Zeist, The Netherlands, 2004
tweede druk, 2e oplage
ISBN 978 90 5011 199 7
www.knnvpublishing.nl

Printed in the United States
By Bookmasters